Regional Assessment of Climate Change in the Mediterranean

ADVANCES IN GLOBAL CHANGE RESEARCH

VOLUME 52

Editor-in-Chief

Martin Beniston, *University of Geneva, Switzerland*

Editorial Advisory Board

B. Allen-Diaz, *Department ESPM-Ecosystem Sciences, University of California, Berkeley, CA, U.S.A.*
R.S. Bradley, *Department of Geosciences, University of Massachusetts, Amherst, MA, U.S.A.*
W. Cramer, *Department of Global Change and Natural Systems, Potsdam Institute for Climate Impact Research, Potsdam, Germany.*
H.F. Diaz, *Climate Diagnostics Center, Oceanic and Atmospheric Research, NOAA, Boulder, CO, U.S.A.*
S. Erkman, *Institute for communication and Analysis of Science and Technology–ICAST, Geneva, Switzerland*
R. Garcia Herrera, *Faculated de Fisicas, Universidad Complutense, Madrid, Spain*
M. Lal, *Center for Atmospheric Sciences, Indian Institute of Technology, New Delhi, India.*
U. Luterbacher, *The Graduate Institute of International Studies, University of Geneva, Geneva, Switzerland.*
I. Noble, *CRC for Greenhouse Accounting and Research School of Biological Science, Australian National University, Canberra, Australia.*
L. Tessier, *Institut Mediterranéen d'Ecologie et Paléoécologie, Marseille, France.*
F. Toth, *International Institute for Applied Systems Analysis Laxenburg, Austria.*
M.M. Verstraete, *Institute for Environment and Sustainability, Ec Joint Research Centre, Ispra (VA), Italy.*

For further volumes:
http://www.springer.com/series/5588

Antonio Navarra • Laurence Tubiana
Editors

Regional Assessment of Climate Change in the Mediterranean

Volume 3: Case Studies

Editors
Dr. Antonio Navarra
National Institute of Geophysics
  and Volcanology
via di Vigna Murata, 605
00143 Rome, Italy

Prof. Laurence Tubiana
IDDRI SciencesPo
41, rue du Four
75006 Paris, France

Euro-Mediterranean Center on Climate
  Change (CMCC)
via Augusto Imperatore, 16
73100 Lecce
Italy

ISSN 1574-0919
ISBN 978-94-007-5768-4   ISBN 978-94-007-5769-1 (eBook)
DOI 10.1007/978-94-007-5769-1
Springer Dordrecht Heidelberg New York London
©Manfred Steinbach/Getty Images/iStockphoto

Library of Congress Control Number: 2013932476

© Springer Science+Business Media Dordrecht 2013
This work is subject to copyright. All rights are reserved by the Publisher, whether the whole or part of the material is concerned, specifically the rights of translation, reprinting, reuse of illustrations, recitation, broadcasting, reproduction on microfilms or in any other physical way, and transmission or information storage and retrieval, electronic adaptation, computer software, or by similar or dissimilar methodology now known or hereafter developed. Exempted from this legal reservation are brief excerpts in connection with reviews or scholarly analysis or material supplied specifically for the purpose of being entered and executed on a computer system, for exclusive use by the purchaser of the work. Duplication of this publication or parts thereof is permitted only under the provisions of the Copyright Law of the Publisher's location, in its current version, and permission for use must always be obtained from Springer. Permissions for use may be obtained through RightsLink at the Copyright Clearance Center. Violations are liable to prosecution under the respective Copyright Law.
The use of general descriptive names, registered names, trademarks, service marks, etc. in this publication does not imply, even in the absence of a specific statement, that such names are exempt from the relevant protective laws and regulations and therefore free for general use.
While the advice and information in this book are believed to be true and accurate at the date of publication, neither the authors nor the editors nor the publisher can accept any legal responsibility for any errors or omissions that may be made. The publisher makes no warranty, express or implied, with respect to the material contained herein.

Printed on acid-free paper

Springer is part of Springer Science+Business Media (www.springer.com)

CIRCE – Climate Change and Impact Research: The Mediterranean Environment

Foreword

CIRCE, co-ordinated by INGV (Istituto Nazionale di Geofisica e Vulcanologia, Italy) and supported by Sixth Framework Programme of European Commission, confirmed the climate change trends in the Mediterranean area indicated in the past IPCC (Intergovernmental Panel on Climate Change) reports, implementing a new generation of climate models for the next decades. These models can be used for simulations on future climate scenarios that are more realistic and detailed. CIRCE realized for the first time a regional assessment of climate change impacts, targeting one of the most complex, socially and physically diverse region of the world. The coordinate approach showed that climate will significantly impact major economic drivers and affect social dynamics.

Water will become an increasingly precious resource, to be managed both for domestic and industrial usages and for irrigation. New policies will have to be developed to adapt to water scarcity, such as water recycling and innovative global water management approaches. It is a challenge because water management systems are often deeply entrenched in local traditions and cultures and therefore they cannot be changed overnight. Water scarcity, which is usually well managed in European countries, is still an issue in North Africa, and it is a clear example of how climate change could amplify regional inequalities.

Reduced water availability and increased frequency and intensity of heat waves will render ecosystems more vulnerable, since climate change is too fast to allow ecosystems adaptation. Particularly affected are traditional crops (wheat, olives, grapes), both because there is less available time for biomass accumulation and because of higher temperatures and water stress on crops. Forests in the Mediterranean region are also going to be affected. CIRCE has shown that it is extremely important to set up adaptation strategies in this field, such as biodiversity and habitats preservation, sustainable development and methods to control fires and pests.

The Mediterranean is the most important tourist region of the world. Climate change will affect tourism fluxes by decreasing the tendency to travel from northern countries as they take advantage of hotter and longer summers, and in the future the south could be strongly affected by a gradual loss of tourists. The relative differences of temperature and precipitation will shape the tourism patterns of the future, and therefore it is important to develop strategic plans both to support tourism resilience and development in new zones and to reduce carbon emissions related to tourist travel.

The analysis of the results produced by the CIRCE economic models showed that the climate change impact on Gross Domestic Product (GDP) might be a decrease of 1% in 2050, or even more (about 3%) in North Africa and in small Mediterranean islands (Malta, Cyprus), which are expected to be more sensitive to climate change. CIRCE analysis is showing that supporting a 'green economy' can provide employment and sustainable development in the southern Mediterranean, reducing inequalities and creating innovative growth opportunities. In this way, policies that take into account the factor of a changing climate could be the occasion for a long-term social and economic improvement for all the countries in the Mediterranean region.

CIRCE has created a community where physical scientists, ecologist, economist and social scientist have worked together effectively, crossing disciplinary barriers and developing a common language. When we will look back to these years, we will have no trouble in realizing that CIRCE has started a new era in climate studies of the Mediterranean region, showing that it is possible to join knowledge and expertise from far disciplines to produce a coherent and consistent assessment that can be a basis of effective adaptation and mitigation policies.

Minister for the Environment, Land and Sea   Corrado Clini
Italy

CIRCE – Climate Change and Impact Research: The Mediterranean Environment

Preface

The CIRCE project started on 1 April 2007 and ended on 30 June 2011, coordinated by the Istituto Nazionale di Geofisca e Vulcanologia, Italy. The work of the CIRCE project was split into 13 research lines plus one for coordination and communication. Research lines were subdivided into several work packages activities.

The main objectives of CIRCE were to predict and to quantify the physical impacts of climate change in the Mediterranean, and to assess the most influential consequences for the population of the region. The knowledge yielded by the different specialized investigations were linked in an integrated interdisciplinary approach designed to study the total effect of climate change. CIRCE integrated cutting-edge scientific research with the needs of end-users and communities. Thus, CIRCE was able to quantify the impact of global warming on Mediterranean climate variables, while also taking into account the regional, social, economic and policy aspects of the process. That allowed to overcome the imbalance faced by a number of research projects on climate impacts between physical and natural science and social sciences so as to build a new vision of the interactions between climate factors and socioeconomic evolutions.

The impacts of climate change were analyzed and evaluated in their oceanographic, meteorological, ecological, economic and societal dimensions, and the project focused particularly on the direct economic impacts for six crucial sectors for the Mediterranean region: health, tourism, energy demand, agriculture, water and human migration. In this way, CIRCE made a powerful contribution to the definition and evaluation of adaptation and mitigation strategies.

The end products of CIRCE have been published in the open scientific literature and summarized in less technical terms in the final report – Regional Assessment of Climate Change in the Mediterranean (RACCM), a decision support system tool for adaptation and mitigation strategies tailored specifically for the Mediterranean environment.

This report represents the synthesis of the work in the project and also an opportunity for an assessment of the present state of the science results for climate change studies and their impacts in the Mediterranean region. Assessments of climate change and their impact at global level have been performed repeatedly in recent years. The methodology and the process have been established during four reports prepared by the Intergovernmental Panel on Climate Change (IPCC) since 1990 and a fifth global report is under preparation.

The response of the climate system to external perturbation, like the increase of greenhouse gases, is highly variable in space and time. The intricate web of nonlinear processes and interactions that link the various components of the system modulates and modifies the overall trend and average response with an intrinsic variability that can be of the same magnitude as the signal from the greenhouse gases. The tools at our disposal to produce quantitative evaluation and estimation, mostly models, have been continuously developing and they are now reaching a level where that can actually provide information at scales smaller than the traditional global scale. On the other hand, as the full extent of possible climate change on our societies and economies was becoming clear, the interest for climate change impact informations at local and regional level has been steadily growing. Global assessments like those pioneered by the IPCC need to be complemented by Regional assessments that take into account the specific climatological, social and economic aspects of a region. Such assessments are now greatly needed to provide a sound scientific base to draft detailed adaptation strategies, inform policies and evaluate their costs.

The progress of the regional assessments has been slowed by the absence of tools adequate to address regional and local issues. CIRCE is pioneering the development of models tailored for the regional climate, either using limited area models or high resolution global models, taking great care to properly represent processes and dynamics that are particularly relevant for the region under study, like for instance an accurate representation of the dynamics of the Mediterranean Sea. Such models allow the delivery of primary climate drivers to the impact studies that reflect more accurately the specific characteristics of the region and therefore should allow in principle a more accurate analysis of the downstream impacts.

The Mediterranean Sea area is a critical area for political, social and economical reasons, but it is also a critical area from the climate point of view. It has a unique position at the border of the tropical zones and the mid-latitude areas, resulting in a complex interplay of interactions between the mid-latitude atmospheric dynamics and tropical processes. The delicate energy and hydrological balance of the Mediterranean Sea influence the Atlantic circulation and, ultimately, the world ocean circulation. The region has unique vulnerabilities from the climate and societal point of view that can go critical under climate change. And yet, the region has been rather under-investigated in recent years and comprehensive analyses and assessments of the region are rare. This report provides the first comprehensive assessments of climate change and its impacts in the Mediterranean region, covering different sectors, from physical climate drivers as temperature and precipitation, to agriculture, forests, from water resources to social impacts, evaluating policies and determining costs of actions and inaction.

The report contains novel results obtained by the new system developed under CIRCE integrating them with some of the existing body of results for the region to achieve an overall evaluation of the state of scientific knowledge regarding climate change and its impacts in the Mediterranean region.

The report is divided in five parts. Parts I and II are collected in the first volume, Parts III and IV made up the second volume and the case studies are presented in the third volume. Though there is a logical sequence to the volumes, the discussion is sufficiently independent that each volume can be considered on its own.

Part I deals with climate change in the atmosphere and the Mediterranean Sea and it contains six chapters: Chapter 1 *Introduction*. Chapter 2 *Past and current climate changes in the Mediterranean Region,* assess the current level of knowledge of the observed climate variability and trends in the Mediterranean, and it includes description of available temperature and precipitation station and gridded data sets, reviewing issues linked to quality control harmonization and homogenization of data; data for the ocean circulation, sea level and waves are also discussed as the changes in extreme events. Chapter 4 *Mechanisms of Climate Variability, Air Quality and Impacts of Atmospheric Constituents in the Mediterranean Region Note: this chapter must be placed after chapter 3 Future Climate Projections* which discusses regional patterns and variability linking them with air quality and direct and indirect impacts on regional climate and associated uncertainties. Chapter 3 *Future Climate Projections,* discusses the status of the art of model projections with existing models and with the new CIRCE models and simulations, including an assessment of the uncertainties of the model projections. Chapter 5 *Detection and Attribution* discusses the issues connected with the identification of the climate change signal and its causes and Chapter 6 *Summary and Major Findings* concludes the part.

Part II is dedicated to the impacts of climate change on the hydrological cycle of the region. It contains five chapters: Chapter 7 *Introduction*. Chapter 8 *The Hydrological Cycle of the Mediterranean* discusses the hydrological cycle in the Mediterranean region, discussing the characteristic and mechanism of the hydrological cycle using both observations and models to analyze possible impacts on the water as a resource, also including some preliminary discussion of adaptation strategies. The particularly relevant impact of climate change on fresh water bodies is discussed in Chapter 9 *Impacts of Climate Change on Freshwater Bodies: Quantitative Aspects* whereas Chapter 10 *Impacts of Climate Change on Water Quality* deals with the impact of climate change on lakes and how nutrients load in rivers are affected. Part II is concluded by Chapter 11 *Summary and Major Findings.*

Part III is devoted to ecosystem services and it is divided in eight chapters: Chapter 1 *Introduction*. Chapter 2 *Vulnerability of Ecosystem Services in the Mediterranean Region to Climate Changes in Combination with Other Pressures* discusses the ecosystems and ecosystem services projected climatic changes and impacts in the Mediterranean, including an analysis of land use changes and a vulnerability assessment. Chapter 3 *Impact of Climate Variability and Extremes on the Carbon Cycle of the Mediterranean Region* discusses observed impacts of climate variability and extremes on the carbon cycle in the Mediterranean. Chapter 4 *Climate Change Impacts on Typical Mediterranean Crops and Evaluation of Adaptation Strategies* analyzes the impacts on selected Mediterranean crops, and the impacts on forest are discussed in Chapter 5 *Climate Change Impacts on Forests and Forest Products in the Mediterranean Area*. Chapter 6 *Effects of Climate and*

Extreme Events on Wildfire Regime and Their Ecological Impacts discusses the impacts of climate change on wildfires regimes. Chapter 7 *Climate Induced Effects on Livestock Population and Productivity in the Mediterranean Area* discusses the effect of climate change on livestock, and finally Chapter 8 *Summary and Major Findings* presents the conclusion.

Part IV contains the analyses on socio-economic impacts of climate change in the region and it is composed of ten chapters: Chapter 9 *Introduction*. Chapter 10 *Integrated Socio-Economic Assessment (The Economic Point of View)* contains an economic assessment of climate change impacts for the Mediterranean region. Chapter 11 *Water and People: Assessing Policy Priorities for Climate Change Adaptation in the Mediterranean* discusses the challenge to water resources with some policy options analyzed through an analysis of the adaptation capacity. General adaptation issues are faced in Chapter 12 *Adaptation Strategies for the Mediterranean*, which discusses how science and scientific results can be used as an input to adaptation strategies, including consideration of uncertainties. Chapter 13 *Health* is devoted to assessing health effects and developing adaptation strategies, whereas energy issues are discussed in Chapter 14 *Energy Demand and GHG Mitigation Options*, which discusses the impacts of climate change on energy markets, discussing options for mitigation and adaptation. The major role of the tourist industry in the Mediterranean region is taken over by Chapter 15 *Mediterranean Tourism and Climate Change: Identifying Future Demand and Assessing Destinations Vulnerability*, which analyzes how climate change would threaten coastal tourism and the expected consequences of climate change. The issue of migration and how climate change can affect the migratory fluxes of retirement is discussed in Chapter 16 *International Retirement Migration from Northern Europe to the Mediterranean: New Results on the Role of Climate with a Possible Application to Climate Change*. Chapter 17 *Green Growth in the Mediterranean* is an essay on sustainable development path in the Mediterranean. And Chapter 18 *Summary and Major Findings* will close the part.

Part V is dedicated to the results from the case studies that have been performed in CIRCE. They provide interesting additional materials to the assessments in the other parts, and certainly they represent a rare example of integrated analysis of the impact of climate change. The case studies have been chosen to represent urban, rural and coastal environments, drawn from the north and the south of the Mediterranean shore. The part discusses also stakeholders involvement, the level of engagement and the data and knowledge indicators used in the assessment and adaptation strategies.

We are confident that this report will be a significant contribution to the advancement of understanding and knowledge in climate change and its impact in the Mediterranean and it will be a useful contribution to the coming IPCC AR5 IPCC report, but it will also be able to inform national and European policies, in particular the follow-up to the White Paper on adaptation to climate change and the development of common European policies. Its value will also stretch in time to contribute to the path towards the 2015 UNFCCC Review.

Antonio Navarra and Laurence Tubiana

Acknowledgments

The research leading to these results has received funding from the 6th Framework Programme (FP6/2002–2006) under grant agreement number 360961 – Project 'Climate Change and Impact Research: the Mediterranean Environment (CIRCE)' whose results have given birth to the 'Regional Assessment of Climate Change in the Mediterranean'.

Intellectual guidance and support over the whole life of this demanding project has been provided by the Project Officers of the European Commission, Georgios T. Amanatidis, Henrietta Hampel, Gerardus Klaassen, Elisabeth Lipiatou, Philippe Quevauviller, Andreas Tilche as well as by the CIRCE Advisory Board composed by Valerio Calzolaio, Juan Carlos Ciscar, Klaus Hasselmann, Martin König, Fedor Mesinger, and Evangelos Papathanassiou.

Furthermore, we thank companies CLU Srl and Zadig SpA, the CIRCE Project Office formed by Lorella Reda and Ilaria Mazzocco and the CIRCE Communication Office for their constant and precious support during the execution of the project.

A number of people contributed to the development of the 'Regional Assessment of Climate Change in the Mediterranean' and deserve recognition.

Firstly, we kindly thank the chapter authors for their contributions, whose work on a wide range of scientific and regional case studies form the substantive basis of this book.

Manuscript reviewers Asbjorn Aaheim, Pascale Delecluse, Klaus Hasselmann, Martin König, John Latahm, María Máñez Costa, Fedor Mesinger, Andy Morse, Reimund Schwarze, and George Tselioudis provided thoughtful and high valuable comments that have contributed to the quality, clarity, and scientific rigor of the information herein.

The managing editors for this book, Margaret Deignan and Takeesha Moerland-Torpey, have been highly supportive and efficient during the book development and were a pleasure to work with.

Centro Euro-Mediterraneo sui Cambiamenti Climatici   Antonio Navarra
Istituto Nazionale di Geofisica e Vulcanologia

Institut du développement durable et des   Laurence Tubiana
relations internationales

Contents

Part I Case Studies

1 Introduction 3
Maureen D. Agnew, Clare M. Goodess, Debbie Hemming,
Christos Giannakopoulos, Skander Ben Salem, Marco Bindi,
Mohamed Nejmeddine Bradai, Letizia Congedi, Camilla Dibari,
Hesham El-Askary, Mutasem El-Fadel, Roberto Ferrise,
José M. Grünzweig, Ali Harzallah, Abdallah Hattour, Maria Hatzaki,
Dina Kanas, Effie Kostopoulou, Piero Lionello, Theib Oweis,
Cosimo Pino, Marco Reale, Agustín Sánchez-Arcilla,
and Mohamed Senouci

2 Stakeholders 23
Maureen D. Agnew, Clare M. Goodess, Debbie Hemming,
Christos Giannakopoulos, Marco Bindi, Camilla Dibari,
Hesham El-Askary, Mamdouh El-Hattab, Mohamed El-Raey,
Roberto Ferrise, Ali Harzallah, Maria Hatzaki, Effie Kostopoulou,
Piero Lionello, Salah Sahabi Abed, Agustín Sánchez-Arcilla,
Mohamed Senouci, Rolf Sommer, Mohamed Zoheir Taleb,
and Annalisa Tanzarella

3 Physical and Socio-economic Indicators 39
Maureen D. Agnew, Clare M. Goodess, Debbie Hemming,
Christos Giannakopoulos, Skander Ben Salem, Marco Bindi,
Mohamed Nejmeddine Bradai, Camilla Dibari, Hesham El-Askary,
Mutasem El-Fadel, Mohamed El-Raey, Roberto Ferrise,
José M. Grünzweig, Ali Harzallah, Abdallah Hattour, Maria Hatzaki,
Dina Kanas, Effie Kostopoulou, Piero Lionello, Theib Oweis,
Cosimo Pino, Basil Psiloglou, Salah Sahabi Abed,
Agustín Sánchez-Arcilla, Mohamed Senouci, Mohamed Zoheir Taleb,
Annalisa Tanzarella

| 4 | **Climate Impact Assessments**... | 61 |

Debbie Hemming, Maureen D. Agnew, Clare M. Goodess,
Christos Giannakopoulos, Skander Ben Salem, Marco Bindi,
Mohamed Nejmeddine Bradai, Letizia Congedi, Camilla Dibari,
Hesham El-Askary, Mutasem El-Fadel, Mohamed El-Raey,
Roberto Ferrise, José M. Grünzweig, Ali Harzallah, Abdallah Hattour,
Maria Hatzaki, Dina Kanas, Piero Lionello, Mark McCarthy,
César Mösso Aranda, Theib Oweis, Joan Pau Sierra,
Basil Psiloglou, Marco Reale, Agustín Sánchez-Arcilla,
Mohamed Senouci, and Annalisa Tanzarella

| 5 | **Integration of the Climate Impact Assessments with Future Projections** .... | 105 |

Clare M. Goodess, Maureen D. Agnew, Christos Giannakopoulos,
Debbie Hemming, Skander Ben Salem, Marco Bindi,
Mohamed Nejmeddine Bradai, Letizia Congedi, Camilla Dibari,
Hesham El-Askary, Mutasem El-Fadel, Mohamed El-Raey,
Roberto Ferrise, Dimitra Founda, José M. Grünzweig, Ali Harzallah,
Maria Hatzaki, Gillian Kay, Piero Lionello, César Mösso Aranda,
Theib Oweis, Joan Pau Sierra, Basil Psiloglou, Marco Reale,
Agustín Sánchez-Arcilla, Mohamed Senouci, Annalisa Tanzarella,
and Konstantinos V. Varotsos

| 6 | **Synthesis and the Assessment of Adaptation Measures** .... | 163 |

Clare M. Goodess, Maureen D. Agnew, Debbie Hemming,
Christos Giannakopoulos, Marco Bindi, Camilla Dibari,
Hesham El-Askary, Mutasem El-Fadel, Mamdouh El Hattab,
Mohamed El-Raey, Roberto Ferrise, José M. Grünzweig,
Ali Harzallah, Dina Kanas, Piero Lionello, César Mösso Aranda,
Theib Oweis, Joan Pau Sierra, Marco Reale, Agustín Sánchez-Arcilla,
Mohamed Senouci, Rolf Sommer, and Annalisa Tanzarella

| 7 | **Executive Summary** .... | 203 |

Clare M. Goodess, Maureen D. Agnew, Debbie Hemming,
and Christos Giannakopoulos

Appendices 211

Index 221

List of Box

Box 1.1	Definitions of Key Concepts	9
Box 2.1	Good Practice Checklist for Facilitating Stakeholder Involvement in Research Projects	35
Box 4.1	Urban Heat Island Modeling for CIRCE Urban Case Studies	64
Box 4.2	Carbon Sequestration in the Judean Foothills	76
Box 5.1	The Exceptionally Hot Summer of 2007 in Athens and Its Consequences	114

List of Figures

Fig. 1.1	Schematic framework showing cross-linkages between the integrating case studies and other CIRCE research groups. 'Climate Dynamics' is covered in Part I of the *Regional Assessment of Climate Change in the Mediterranean* (RACCM); 'Impacts' is covered in Parts II and III of the RACCM; 'Societal Dynamics' and 'Induced Policies' is covered in Part IV of the RACCM; and 'Case Studies' in (this) Part V of the RACCM	6
Fig. 1.2	Schematic representation of the CIRCE Case studies Integrating Framework (CCIF)	8
Fig. 1.3	Location map of the three urban case studies: Athens (Greece), Beirut (Lebanon) and Alexandria (Egypt)	12
Fig. 1.4	Location of the four rural case studies: Tuscany and Apulia (Italy), Tel Hadya (Syria), and the Judean Foothills (Israel)	14
Fig. 1.5	Location of the four coastal case studies: the Gulf of Valencia (Spain), the Gulf of Oran (Algeria), the Gulf of Gabès (Tunisia), and the West Nile Delta (Egypt)	17
Fig. 2.1	Ladder of stakeholder participation	27
Fig. 2.2	Vulnerability to coastal flooding by land-use type in Alexandria and the surrounding region, West Nile Delta, Egypt. High spatial resolution digital elevation map in meters (see legend) was used to create contours of elevation. This elevation layer was overlain with a land cover map within a Geographical Information System to identify zones at high risk from coastal flooding	29
Fig. 3.1	Beirut Summer maximum temperature (Tx) anomalies (°C) from the 1971 to 2000 average (*left axis*) for the Aéroport International de Beyrouth (AIB) station (*bars*). 10-year moving average of the absolute summer Tx (*right axis*) for the AIB station (*line*)	48

Fig. 3.2	Annual evapotranspiration (mm/year) in Tuscany for the years 1999 (average precipitation), 2003 (dry year), 2008 (wet year)	49
Fig. 3.3	Administrative municipalities most affected (*deepest red*) in terms of total area burnt (*left*). Fire Weather Index (*red line*) and number of summer fire events (blue bars) from 1984 to 2005 (*right*)	49
Fig. 3.4	Indicator linkages for the Judean Foothills. The *blue box* denotes key climate drivers, the *green boxes* denote key biogeophysical impacts and vulnerabilities (*italics*); the *orange boxes* denote key social impacts and vulnerabilities (*italics*)..	50
Fig. 3.5	Mean growing season (October to June) daily maximum, minimum and mean temperature (°C) for Tel Hadya, Aleppo-Syria, for the period 1978/1979–2006/2007	51
Fig. 3.6	MODIS satellite sensor derived level 2 (10×10 km) daily aerosol optical depth (AOD) over Alexandria, Egypt during the period 2000–2009..	53
Fig. 3.7	Cumulated number of observed alien species (logarithm) in the Gulf of Gabès per observing year 1966–2006; the *solid line* is the linear trend. First records are based on a compilation of the observed alien species (40) from available papers, thesis and reports. The *inner plot* shows the increasing number of studies on alien species (to remove the bias introduced by the increased research, and hence reporting, activity)	55
Fig. 3.8	Daily excess summer deaths in Athens by maximum air temperature interval (*blue bars*) for the years 1992–2006. The frequency of occurrence of each temperature interval is shown using *yellow bars*..	58
Fig. 3.9	Water consumption (million m^3) in Tuscany by agriculture (2000–2008), civil (2006 only) and industry (1995–2007) sectors..	58
Fig. 4.1	Frequency of hot nights and days for a present (1971–1990) and a future period (2041–2060) for rural and urban areas, including an additional driver of urban climate change from local anthropogenic heat release (Urban +15 Wm^2, Urban +45 Wm^2)..	65
Fig. 4.2	Summary of key climate-related impacts at the three CIRCE urban case study locations together with the associated hazards and vulnerabilities (both social and biogeophysical). *Arrows* represent interactions between these categories via climate trends and dynamics and biogeophysical and social dynamics	66
Fig. 4.3	Scatter plot of daily energy consumption and daily mean air temperature in Athens, 1997–2001	67

List of Figures   xix

Fig. 4.4   Number of hot days (defined using the 90th percentile
            of daily Tx 1961–1990) shown as anomalies from the 1971
            to 2000 average for Athens, Beirut and Alexandria 70
Fig. 4.5   Total annual rainfall shown as anomalies (mm) from the
            1971–2000 average for Athens, Beirut and Alexandria 71
Fig. 4.6   Summary of key climate-related impacts (*green*) at the three
            CIRCE rural case study locations together with the associated
            hazards (*blue*) and vulnerabilities (*orange*) (both social
            and biogeophysical). Arrows represent interactions between
            these categories via climate trends and dynamics, and
            biogeophysical and social dynamics... 74
Fig. 4.7   Observed maximum and minimum temperature 1951–2000
            for the Apulia case study site... 81
Fig. 4.8   Wheat grain yield (t/ha) of various rain-fed varieties
            at the farm level relative to total annual precipitation
            (mm) in two regions of the Judean Foothills................................ 83
Fig. 4.9   Groundwater levels in Tel Hadya, 1983–2007 84
Fig. 4.10   Stem volume (m³/ha) of Pinus halepensis in planted
            forests (n = 14) along a rainfall gradient. Forests stands
            were aged 35–45 year with a density of 250–350 trees/ha.
            The linear regression is significant at p = 0.005 84
Fig. 4.11   Summary of key climate-related Impacts (*green*) at the four
            CIRCE coastal case study locations together with the
            associated hazards (*blue*) and vulnerabilities (*orange*)
            (both social and biogeophysical). *Arrows* represent interactions
            between these categories via climate trends and dynamics,
            biogeophysical and social dynamics, and hazard/vulnerability
            feedbacks 90
Fig. 4.12   Annual mean sea level (cm) at the Ebro Delta, 1958–2001
            (HIPOCAS hindcast model data) together with the slope
            estimate (*red line*) with the 95% confidence interval
            (*blue dashed lines*).. 93
Fig. 4.13   Yearly maximum hindcast significant wave height values (m)
            for the Catalan coast (1959–2001) 93
Fig. 4.14   Annual maximum significant swell height (m) 1974–1993
            in three illustrative locations of the Gulf of Gabès 94
Fig. 4.15   Annual distribution of storms in Oran during the period
            1950–2007. A storm event is defined according to the World
            Meteorological Organisation observation criteria of lightning
            and thunder storms reported at the same station 94
Fig. 4.16   Annual mean (*yellow*), maximum (*red*) and minimum (*blue*)
            sea surface temperature (SST) at the Northern Catalan Coast
            (Estartit), 1969–2008 (*top*) and, (b) air temperature evolution,
            1905–2008 (*bottom*) 95

Fig. 4.17 Mean annual maximum temperature anomalies (from 1971 to 2000 average) for Gulf of Oran for 1927–2007 (*top left*) and West Nile Delta for 1973–2009 (*top right*). Middle figures show number of hot days, where maximum temperature exceeds the 1961–1990 90th percentile temperatures for Gulf of Oran (*left*) and West Nile Delta (*right*). *Bottom figures* show monthly anomalies of air temperature in the Gabès station for 1948–2008 based on data archived at the Goddard Institute for Space Studies (GISS, www.giss.nasa.gov). Trends with 95% confidence limits are shown for the period 1948–2008 and 1973–2008. The *bottom left figure* is annual and the *bottom right figure* is for the summer season only 96

Fig. 4.18 Daily Climate Tourism Potential Index time series (1972–2003) for the classes 'favorable days' (a) and 'acceptable days' (b) for Djerba (Gulf of Gabès). *Linear trends* are shown with the trend rates (in % days/decade) 98

Fig. 5.1 Projected changes in annual Tmax (°C) (2021–2050 minus 1961–1990) based on CIRCE model data for the urban, rural and coastal case studies. The ensemble-mean change is shown, together with changes for the five individual models. IPSL is IPSLreg 111

Fig. 5.2 Projected changes in Tmin (°C) (2021–2050 minus 1961–1990) based on CIRCE model data for the urban, rural and coastal case studies. The ensemble-mean change is shown, together with changes for the five individual models. IPSL is IPSLreg 111

Fig. 5.3 Projected changes (2021–2050 minus 1961–1990) in the number of very hot summer days (Tx95n) based on CIRCE model data for the urban, rural and coastal case studies. The ensemble-mean change is shown, together with changes for the five individual models. IPSL is IPSLreg 112

Fig. 5.4 Projected seasonal changes (2021–2050 minus 1961–1990) in the number of very hot nights (Tn95n) based on CIRCE model data for the urban case studies. The ensemble-mean change is shown, together with changes for the five individual models. IPSL is IPSLreg 113

Fig. 5.5 Probability density function (pdf) of Gaussian distributions fitted to Athens JJA maximum (*left*) and minimum (*right*) temperature for the following cases: (*a – solid black*) model output for 1961–1990, (*b – dashed black*) NOA observations for 1961–1990 (*c – green*) 2021–2050 model simulations, (*d – dashed red*) 2071–2100 model simulations and (*e – solid red*) summer 2007 (NOA observations) 115

Fig. 5.6	Projected changes (2021–2050 minus 1961–1990) in summer WSDI (Warm Spell Duration Index: maximum number of consecutive days (at least 6 days) with Tmax > 90th percentile) for five CIRCE models (CNRM, ENEA, INGV, IPSLreg, MPI)	117
Fig. 5.7	Projected seasonal changes (2021–2050 minus 1961–1990) in the number of very cold nights (Tn5n) based on CIRCE model data for the rural case studies. The ensemble-mean change is shown, together with changes for the five individual models. IPSL is IPSLreg	118
Fig. 5.8	Projected changes (in mm for 2021–2050 minus 1961–1990) in (**a**) annual and (**b**) extended winter (October to March) total precipitation based on CIRCE model data for the urban, rural and coastal case studies. The ensemble-mean change is shown, together with changes for the five individual models. IPSL is IPSLreg	119
Fig. 5.9	Projected changes in three precipitation extremes indices for the urban case studies calculated using daily output from the ENEA and IPSLglo models: the maximum dry spell length (consecutive dry days: cdd), heavy precipitation (the 90th percentile of daily precipitation: pq90) and maximum 3-day precipitation (px3d). A threshold of >0.5 mm was used to define a rain day	120
Fig. 5.10	Simulated sea surface temperatures for 1950–2050 in the Gulf of Gabès from CIRCE simulations. The *thick black line* is the ensemble mean	121
Fig. 5.11	Mean number of fires per day against FWI (*crosses*) for fires near Athens for the period, 1983–1990. The crosses merge to form a *thick curve*, except at large FWI where data are sparse	132
Fig. 5.12	Projected changes (2021–2050 minus 1961–1990) in the number of days with fire risk FWI>15 (*left*) and extreme fire risk FWI>30 (*right*) based on output from six ENSEMBLES RCM simulations	133
Fig. 5.13	Projected changes (2021–2050 minus 1961–1990) in the number of days with fire risk FWI>15 (*left*) and extreme fire risk FWI>30 (*right*), based on the average of two CIRCE models (ENEA and MPI)	134
Fig. 5.14	Scatter plot of the daily 8-h maximum average ozone concentration vs. daily maximum temperature for Athens, with best fit line (*green*) and 95% confidence limits (*red*)	135
Fig. 5.15	Ozone probability distributions for temperature above 17°C, for two periods, observed (1990–1999) (*green*) and 2021–2050 (*black, red, grey*) for CIRCE models (ENEA, MPI, IPSLglo). Values are plotted at the mid-point concentration of each 5 ppb ozone concentration bin	136

Fig. 5.16	Excess deaths (*right axis; red and dark blue bars* for present and future climate with adaptation, respectively) and daily temperature frequencies (*left axis; light blue bars*) in Athens, for the future periods 2021–2050 (*top*) and 2071–2100 (*bottom*), respectively, for the RACMO2 RCM ENSEMBLES simulation..	137
Fig. 5.17	Excess deaths (*right axis; red, dark royal blue, green,* and *pink bars* for present and future climate with adaptation, respectively) and daily temperature frequencies (*left axis; light blue bars*) in Athens, for the future period 2021–2050 derived from three CIRCE multi-model simulations (ENEA, IPSLglo and MPI)..	138
Fig. 5.18	Variation in Athens daily energy consumption (*left axis, red line*) with mean daily air temperature (*right axis, blue line*) for 2001	139
Fig. 5.19	Athens energy demand (*bars, right axis*) and daily maximum temperature (*triangles, left axis*) for the colder (*left*) and warmer (*right*) halves of the year for 1961–1990, 2021–2050 periods for CIRCE simulations (ENEA, MPI, IPSLglo)	140
Fig. 5.20	Estimated number of observed alien species in the Gulf of Gabès (*red line*) obtained as a fitted exponential function to SST (Fitted SST=42.8*SST-741.6). An exponential fit to observations (*blue line*) with a trend of 15%/decade is also shown. The observations (*blue symbols*) are the cumulative number of new species recorded, and are adjusted to correct for the increasing number of studies from a reference year, 2007...	141
Fig. 5.21	Tourism Seasonal Shift Index (T>23°C): the start and end of the summer season calculated using output from a range of CIRCE models for the period 1950–2050. The ensemble-mean (based on all models except ENEA-ERA40-2) is also shown. Days are counted from 1st March. Prior to calculation, the mean daily temperature series from each model was adjusted to the mean observed temperature at Djerba during the period 1973–2003	142
Fig. 5.22	DCTPI (Daily Climate Tourism Potential Index) time series (1950–2050) for the classes 'favorable days' (*top*) and 'acceptable days' (*bottom*) for the Gulf of Gabès. The shown series are calculated from the output of five CIRCE coupled models, and the ensemble mean	143
Fig. 5.23	A theoretical simulation of socio-economic impacts of days becoming 'acceptable' in the Djerba-Zarzis tourist area for the period 1960–2050, based on the ensemble average of five CIRCE coupled models for the Gulf of Gabès area (the average of the different model variables used to calculate the percent	

	days becoming 'acceptable' are adjusted to that of the observations for the period 1973–2006). Results are shown as annual percentages of 2007 GDP. Projections of the direct and indirect employee loss should be multiplied by 100,000. Economic losses are based on 2007 rates	144
Fig. 5.24	Total population growth rates for the case-study countries	146
Fig. 5.25	Indicator linkages for the Gulf of Gabès, Tunisia. The blue boxes denote key climate and marine drivers, the green boxes denote key biogeophysical impacts and vulnerabilities (*italics*); the orange boxes denote key social impacts and vulnerabilities (*italics*)	149
Fig. 5.26	Marine ecosystem – fishing activity vulnerability linkage diagram for the Gulf of Gabès	150
Fig. 5.27	Time series of mean annual maximum (*left*) and minimum (*right*) temperature (°C) for the urban case studies from observations (*black*) and the ENEA (*red*) and IPSLglo (*green*) CIRCE models. The models run from 1950 to 2050	152
Fig. 5.28	Time series of total annual precipitation (mm) for the urban case studies from observations (*black*) and the ENEA (*red*) and IPSLglo (*green*) CIRCE models. The models run from 1950 to 2050	153
Fig. 5.29	Time series of anomalies (with respect to 1961–1990) in mean annual maximum (*left*) and minimum (*right*) temperature (°C) for the urban case studies from observations (*black*) and the ENEA (*red*) and IPSLglo (*green*) CIRCE models. The models run from 1950 to 2050	156
Fig. 5.30	Time series of anomalies (with respect to 1961–1990) in the number of very hot days (*left*) and very hot nights (*right*) for the urban case studies from observations (*black*) and the ENEA (*red*) and IPSLglo (*green*) CIRCE models. The models run from 1950 to 2050	158

List of Tables

Table 1.1	The integrating Mediterranean case studies	6
Table 2.1	Categories of stakeholders relevant to the generic integrating case studies	26
Table 3.1	Organization of the CIRCE case-study indicators	42
Table 3.2	Changes in the diversity of vegetation types ('+' present; '−' absent) in natural ecosystems with changes in mean annual precipitation	46
Table 4.1	Summary table of biogeophysical and social vulnerability indicators for current climate in Athens, including identified system thresholds	68
Table 4.2	Key biogeophysical and social vulnerabilities identified for the CIRCE urban case study locations, Alexandria (Al), Athens (At) and Beirut (Be)	72
Table 4.3	Key biogeophysical and social vulnerabilities identified for the CIRCE rural case study locations, Apulia (Ap), Judean Foothills (JF), Tel Hadya (TH) and Tuscany (Tu)	82
Table 4.4	Summary of biogeophysical and social vulnerability indicators for current climate in Tuscany	87
Table 4.5	Waste water discharge in the Bay of Oran	92
Table 4.6	Key biogeophysical and social vulnerabilities identified for the CIRCE coastal case study locations, Gulf of Gabès (Ga), Gulf of Oran (Or), Gulf of Valencia (Va) and West Nile Delta (WN)	99
Table 5.1	Summary of the CIRCE models	108
Table 5.2	Statistics of summer maximum and minimum temperature, observed for the period 1961–1990 and summer 2007 (NOA), and simulated for the periods: 1961–1990, 2021–2050 and 2071–2100	115

Table 5.3	Capacity for quantitative or qualitative assessments of the consequences of future climate change for the CIRCE integrated case studies	124
Table 6.1	Key vulnerability indicators identified for the CIRCE integrated case studies	168
Table 6.2	National vulnerability and adaptive capacity indicators for the case-studies	170
Table 6.3	Format and approach for the regional stakeholder workshops	172
Table 6.4	Water adaptation strategies considered in the Beirut urban case study	175
Table 6.5	Examples of case-study adaptation options summarized within the framework of the EU White Paper on Adaptation (2009)	182
Table 6.6	Summary of adaptation activities by case-study country	190
Table 6.7	Summary of observed (last few decades) and projected (2021–2050) changes in mean temperature (Mean T), high temperature extremes (High T) and mean precipitation (Mean P) in the case-study regions	194
Table A2.1	Climate indicators	212
Table A2.2	Marine indicators	214
Table A2.3	Coastal impact and vulnerability indicators	215
Table A2.4	Terrestrial ecosystem impact and vulnerability indicators	215
Table A2.5	Agricultural impact and vulnerability indicators	217
Table A2.6	Human health impact and vulnerability indicators	217
Table A2.7	Demographic vulnerability indicators	218
Table A2.8	Socio-economic impact and vulnerability indicators	218

Part I
Case Studies

Chapter 1
Introduction

Maureen D. Agnew, Clare M. Goodess, Debbie Hemming,
Christos Giannakopoulos, Skander Ben Salem, Marco Bindi,
Mohamed Nejmeddine Bradai, Letizia Congedi, Camilla Dibari,
Hesham El-Askary, Mutasem El-Fadel, Roberto Ferrise, José M. Grünzweig,
Ali Harzallah, Abdallah Hattour, Maria Hatzaki, Dina Kanas, Effie
Kostopoulou, Piero Lionello, Theib Oweis, Cosimo Pino, Marco Reale,
Agustín Sánchez-Arcilla, and Mohamed Senouci

Abstract The heterogeneous nature of the Mediterranean environment, combined with a wide diversity of socio-economic and cultural identities, make this region particularly amenable to integrated research on climate change impacts, vulnerabilities, and adaptive response. Eleven case-study locations have been strategically selected to represent three generic Mediterranean environments (urban, rural and coastal). While each case study location comprises a unique and complex set of climate-related issues, the range and scope of the case studies allows identification

M.D. Agnew (✉) • C.M. Goodess
Climatic Research Unit, School of Environmental Sciences,
University of East Anglia, UEA, Norwich, UK
e-mail: m.agnew@uea.ac.uk

D. Hemming
Climate Impacts Analysis Team, Met Office Hadley Centre, Exeter, Devon, UK

C. Giannakopoulos • M. Hatzaki • E. Kostopoulou
Institute for Environmental Research and Sustainable Development,
National Observatory of Athens, Athens, Greece

M. Bindi • R. Ferrise
Department of Plant, Soil and Environmental Science (DIPSA),
University of Florence, Florence, Italy

Institute for Biometeorology, National Research Council of Italy,
(CNR-IBIMET), Florence, Italy

C. Dibari
Department of Plant, Soil and Environmental Science (DIPSA),
University of Florence, Florence, Italy

S. Ben Salem • M.N. Bradai • A. Harzallah • A. Hattour
Institut National des Sciences et Technologies de la Mer – INSTM,
Salammbô, Tunisia

L. Congedi
Centro Euro Mediterraneo per i Cambiamenti Climatici, CMCC, Lecce, Italy

of common lessons and messages for the wider Mediterranean region. The aim is to perform an integrated assessment of climate impacts, vulnerability and adaptation at a regional to local scale. A risk-based 'bottom up' approach (based on regional stakeholder dialogue) is combined with a 'top down' case-study indicator assessment focused on a common conceptual and methodological framework.

Keywords Climate change • Mediterranean • Integrated assessment • Risk • Vulnerability

1.1 Background to the Mediterranean Case Studies

The long history of human settlement across the Mediterranean region has led to extensive changes in land use and depletion of natural resources (such as fresh water), and has radically changed the Mediterranean landscape (Allen 2003). The heterogeneous nature of the Mediterranean environment, combined with a wide diversity of economic, social and cultural identities, make this region particularly amenable to integrated research on climate change impacts, vulnerabilities, and adaptive response.

H. El-Askary
School of Earth and Environmental Sciences, Schmid College of Science and Technology, Chapman University, Orange, CA, USA

Department of Environmental Sciences, Faculty of Science,
Alexandria University, Alexandria, Egypt

M. El-Fadel
Faculty of Engineering and Architecture, American University of Beirut,
Beirut, Lebanon

J.M. Grünzweig • D. Kanas
Institute of Plant Sciences and Genetics in Agriculture,
The Robert H. Smith Faculty of Agriculture, Food and Environment,
The Hebrew University of Jerusalem, Jerusalem, Israel

P. Lionello
Centro Euro Mediterraneo per i Cambiamenti Climatici, CMCC, Lecce, Italy

Department of Material Science, University of Salento, Lecce, Italy

T. Oweis
International Center for Agricultural Research in the Dry Areas, ICARDA,
Aleppo, Syria

C. Pino • M. Reale
Department of Material Science, University of Salento, Lecce, Italy

A. Sánchez-Arcilla
Laboratori d'Enginyeria Marìtima (LIM/UPC), Universitat Politécnica
de Catalunya, Barcelona, Spain

M. Senouci
Association Recherche Climat et Environnement – ARCE, Oran, Algeria

However, few climate change impact and vulnerability assessments have focused on the Mediterranean region. Limited examples include socio-economic perspectives of climate change (Giupponi and Schechter 2003) and the WWF study of climate change impacts in the Mediterranean resulting from a 2°C global temperature rise (Giannakopoulos et al. 2005, 2009). In 2006, a NATO workshop discussed specific climate-related issues facing the Mediterranean region, highlighting in particular, regional air quality and aerosols (Mellouki and Ravishankara 2007). Regional climate impact assessments in the Mediterranean area have focused on particular themes such as fresh water (Milano 2010), fire risk (Moriondo et al. 2006), tourism (Valls and Sardá 2009), agriculture (Olesen et al. 2007), coasts (Georgas 2003) and marine ecosystems (Gambaiani et al. 2009). The CIRCE case-study work represents the first known regional-to-local scale integrated assessments of biogeophysical and social impacts and vulnerability in both European and non-European countries of the Mediterranean region.

Within the framework of the CIRCE project 11 case-study locations were selected to reflect three generic environments (urban, rural, and coastal), to quantify current and future climate change and to assess the potential consequences for human communities and ecosystems at the regional to local scale.

This third volume of the *Regional Assessment of Climate Change in the Mediterranean* is devoted to the CIRCE Mediterranean case-study integrated assessments. This chapter outlines the context of the integrating case studies, providing information on the main objectives, and describing the assessment framework and methodological stages. The background to each of the case studies (Appendix 1) is described including justification for selection. The importance of communication between stakeholders and research communities, and the challenges this presents, is the focus of Chap. 2. Consideration of the methodological approach and use of thematic indicators for assessing past and future climate change, impacts and vulnerability is the subject of Chap. 3, while Chap. 4 discusses some of the key impacts and vulnerabilities for each of the generic categories of case studies (urban, rural and coastal). Chapter 5 integrates current climate risks with projections for the future and includes a discussion of assessment uncertainties and limitations. This volume concludes with Chap. 6, synthesizing the key case-study results and their interpretive value for identifying critical vulnerabilities and effective adaptation strategies, and providing recommendations for research gaps and needs.

Where possible, the case-study assessments have exploited the specialized and high-spatial resolution information on climate change and its impacts available from other CIRCE research groups (Fig. 1.1). These assessments also feed into and benefit from the policy and adaptation work undertaken within other areas of the project. Stakeholder involvement (focusing on local and regional decision and policy makers) together with the broader case-study outputs have allowed inter-comparisons and contrasts to be drawn in terms of the diverse climate regimes of the Mediterranean, together with the differing political and socio-economic systems. At the same time, the range and scope of the case studies allows identification of common lessons and messages for the whole Mediterranean.

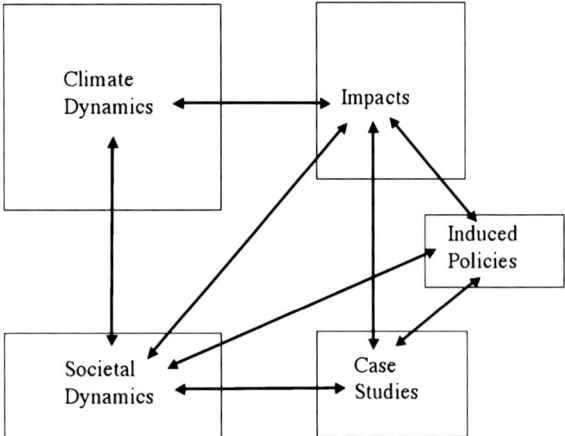

Fig. 1.1 Schematic framework showing cross-linkages between the integrating case studies and other CIRCE research groups. 'Climate Dynamics' is covered in Part I of the *Regional Assessment of Climate Change in the Mediterranean* (RACCM); 'Impacts' is covered in Parts II and III of the RACCM; 'Societal Dynamics' and 'Induced Policies' is covered in Part IV of the RACCM; and 'Case Studies' in (this) Part V of the RACCM

Table 1.1 The integrating Mediterranean case studies

Generic case study	Individual case-study name, country
Urban	Athens, Greece
	Alexandria, Egypt
	Beirut, Lebanon
Rural	Tuscany, Italy
	Apulia, Italy
	Judean Foothills, Israel
	Tel Hadya, Syria
Coastal	Gulf of Valencia – Catalan coast, Spain
	Gulf of Oran, Algeria
	Gulf of Gabès, Tunisia
	West Nile Delta, Egypt

Eleven individual case studies were chosen to reflect the east-west and north-south contrasts across the Mediterranean region and were grouped into three generic types, urban, rural and coastal (Table 1.1). The following criteria were used to select these case-study locations: vulnerability to climate change; availability of appropriate data; and, access to regional stakeholders including decision and policy makers. The issues of importance vary between generic case studies. For example, for rural areas, yields of key crops, water resources, biodiversity and tourism are crucial issues. Whilst for urban areas, energy use, water supply, water quality, water-resource competition between sectors, air quality, thermal comfort, and related health issues, are major concerns. In coastal zones, key research themes involve changes in the marine environment (sea level, wave storms and surges, and sea-water temperature), vulnerability of coastal zones to erosion, flooding and saline intrusion, marine pollution, biodiversity and invasive species, and the corresponding impacts on fisheries and industry (including tourism) located in the coastal zone.

While all case-study locations have sufficient data to allow meaningful assessments to be undertaken, the level of detailed climate and marine data, and the quality and length of data series for biogeophysical and social information vary across sites. In the case of coastal deltas, for example, the scale of physical processes is finer than the spatial resolution of climate models which merely distinguish between land and sea boxes. For urban areas, the global and regional climate models used for projections do not yet incorporate the physics of the urban heat island although some modeling work on this issue has been undertaken as part of the case-studies work (Sect. 4.2; Box 4.1). Such limitations in the current generation of climate models do not negate the need for stakeholders and policy makers to plan for climate change and to identify practical adaptation and mitigation strategies (Wilby et al. 2009). However, it is important in the decision-making process and in the development of climate adaptation strategies to explicitly recognize the existence and nature of uncertainty at all stages of assessment (Sect. 5.4). For this reason, the term climate 'projections' is always used in the case-study work and 'prediction' avoided. At the same time, non-climatic factors are given considerable emphasis, particularly in the context of vulnerability and adaptation. Thus the case-study approach is not incompatible with robust decision making approaches which are considered promising and appropriate in the context of climate change adaptation (Dessai et al. 2009; Part IV, Sect. 12.4.2).

1.2 Objectives

The case-studies integrated assessments had four key objectives:

1. To integrate state-of-the-art climate and impacts modeling studies focused on the Mediterranean
2. To use a rigorous common framework, including a set of quantitative environmental, vulnerability and sustainability indicators developed specifically for the Mediterranean environment
3. To use a risk-based bottom-up, iterative approach (based on regional stakeholder workshops), as well as a top-down approach (based on indicator assessment) to identify and explore potential adaptation and mitigation measures
4. To consider North African (Algeria, Tunisia, Egypt), Middle Eastern (Israel, Lebanon, Syria) and European (Spain, Italy, Greece) case-study locations.

These objectives were largely accomplished through the course of the project, although limitations of time and financial resources, and a dependence on available data and knowledge have in places constrained or necessitated slight changes to the original research strategy.

1.3 The Case-Studies Integrating Framework

A rigorous common framework, referred to as the CIRCE Case studies Integrating Framework – CCIF (Fig. 1.2), was developed to assess the integrated risks of climate change for the selected Mediterranean case-study locations. The use of a conceptual framework facilitates a structured and systematic basis for identifying and selecting indicators (Niemeijer and de Groot 2008). Within this framework, climate dynamics is viewed as a key driver of change in social and biogeophysical environments and is modulated by the inherent dynamics of these systems. The frequency and magnitude of climate hazards is altered by changes in the climate state and has measurable impacts on physical and social systems. The outcomes can be complex, resulting from direct and indirect effects of several climate and non-climate factors. Furthermore, the level of impact is altered by the system vulnerability to climate

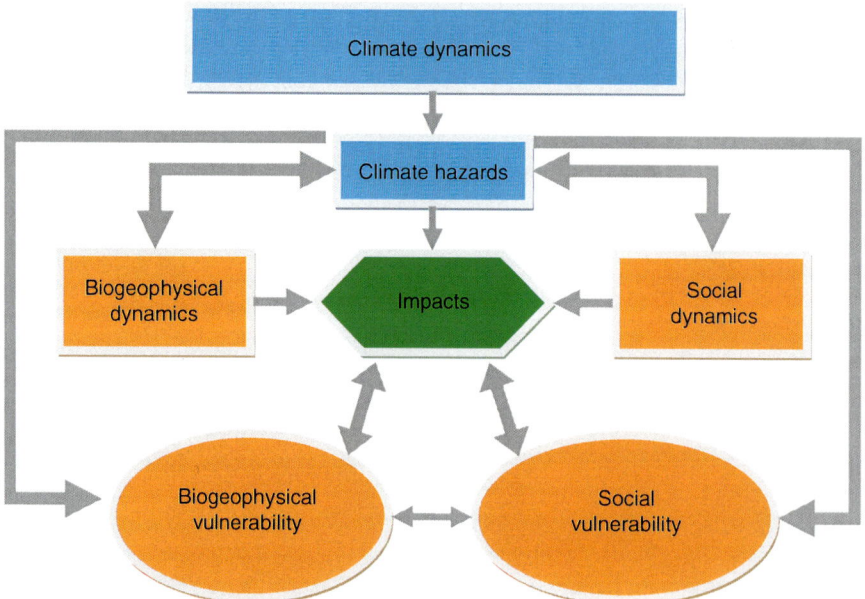

Fig. 1.2 Schematic representation of the CIRCE Case studies Integrating Framework (CCIF)

variability and change, and the risk involved is determined by the probability of the hazard occurring. Societal and environmental vulnerability to climate change is a function of the degree of exposure, the sensitivity of the system, and the capacity for adaptation. Box 1.1 defines the key terms included in the CCIF.

Box 1.1 Definitions of Key Concepts

The following definitions refer to the key elements of the CCIF represented in Fig. 1.2, and largely adhere to those definitions given in the Working Group II Fourth Assessment Report (AR4) of the Intergovernmental Panel on Climate Change (IPCC 2007).

Climate dynamics refer to variability and change in the climate system including for example, changes in temperature, precipitation, solar radiation and cloud cover, large-scale circulation patterns, wind strength and direction.

Climate hazard can be any potentially damaging event or change in climate, such as a single sudden onset extreme event that exceeds a critical temperature threshold, or a complex combination of longer-term changes involving multiple climate variables and/or resulting in multiple impacts (Brooks 2003; Brooks et al. 2005; UNDP 2005). To determine the risk involved with a particular hazard, it is necessary to consider the likelihood or probability of its occurrence at a particular location over a certain period of time (Taubenböck et al. 2008). The risks of climate change or climate hazards are typically defined by criteria (usually thresholds) that link the impacts of climate change to their potential outcomes (Carter et al. 2007). These thresholds can be defined through research or through stakeholder consultation (Conde and Lonsdale 2005), and they contribute to the development of a vulnerability framework.

Impacts are measurable outcomes of (or system responses to) climate dynamics and climate hazards, and are typically modulated by changes in biogeophysical and social systems. Impact categories covered by the CIRCE case studies include: health; tourism; agriculture; biodiversity; water; and energy. Some measurable outcomes, for example, water quality, coastal erosion, and invasive species, are also vulnerability indicators for other measurable outcomes (e.g., marine biodiversity; coastal tourism and fisheries) in a cascading system of impacts.

Biogeophysical dynamics describe change and variability in the biological, geochemical, and physical environmental systems (other than climate), such as air, soil, and water quality; sea level; biodiversity; fish stocks; and land use.

(continued)

> **Box 1.1** (continued)
>
> **Social dynamics** describe changes in social, economic and political systems, such as demography (population growth, structure and density), morbidity, migration, education, technology, finance, civil stability and governance.
>
> **Vulnerability** has been defined in a variety of ways in the climate change literature (e.g., Downing and Patwardhan 2005; Adger 2006; Füssel and Klein 2006). Common to many definitions is the concept that vulnerability is a function of the exposure and sensitivity of a system to a climate hazard, and the ability to adapt to the hazard. Here, the definition developed by Working Group II of the Fourth Assessment Report of the Intergovernmental Panel on Climate Change (IPCC AR4) has been adopted:
>
> 'Vulnerability is the degree to which a system is susceptible to, and unable to cope with, adverse effects of climate change, including climate variability and extremes. Vulnerability is a function of the character, magnitude, and rate of climate change and variation to which a system is exposed, its sensitivity, and its adaptive capacity' (IPCC 2007).
>
> Both biogeophysical vulnerabilities (for example, water scarcity, soil erosion, and marine pollution), and social vulnerabilities (for example, high population density, poverty, and human disease) have been considered.

Central to the case-studies integrated assessment, was the development and use of a set of climate change vulnerability and impact indicators, stakeholder participation, and the consideration of adaptation measures. For each case-study assessment, a common set of methodological steps was followed:

- Identify key climate issues and sensitivities through literature ('top-down' approach) and in dialogue with an appropriate community of local and regional stakeholders ('bottom-up' approach).
- Design a case-study specific conceptual framework (based on the CCIF) that describes linkages between the key climate hazards, impact sectors and biogeophysical and social vulnerabilities.
- Compile a set of indicators for specific climate hazards (present and future) using the best available observed and modeled data sets, sector impacts, biogeophysical and social vulnerabilities. This should involve an iterative approach with several rounds of consultation and review.
- Perform a risk-based integrative assessment of cross-sector impacts of climate hazards using case-study sets of indicators. For an individual case study, this integrated assessment may be of a more qualitative than quantitative nature depending on the nature and availability of data and information.
- Identify any critical thresholds in the system response to climate dynamics using observed data, previous research, and expert knowledge.

- In collaboration with stakeholders and other project research groups (considering socio-political dynamics), identify and evaluate the effectiveness of adaptation and, if possible, mitigation options.
- Provide policy-relevant advice based on the study outcomes.

1.4 Case Studies

1.4.1 Introduction

The Mediterranean Sea connects the coastlines of countries in Europe, the Middle East and North Africa, engendering the region environmentally and culturally unique and diverse. Eleven case studies were strategically chosen to provide valuable opportunities for local and regional stakeholder involvement, and to allow adaptation issues to be addressed using both bottom-up and top-down approaches. The CCIF approach (Sect. 1.3) was applied to each of the case studies, providing a common methodological framework for an indicator-based assessment of the cross-sector impacts of climate change.

Each of the case studies is briefly outlined in the following sections, highlighting the areas of vulnerability (one of the case-study selection criteria – see Sect. 1.1) identified at the project proposal stage and during the early stages of the project. In some cases, the focus of the case study was modified during the course of the project, reflecting discussions with local stakeholders (Chap. 2) and the processes of indicator selection (Chap. 3), construction and analysis (Chaps. 4, 5 and 6).

1.4.2 Urban

Many of the cities and settlements of the ancient world were founded along the Mediterranean coastline and nearby river basins which provided the necessary resources to meet the needs of early civilizations. Today these cities, especially those in the east and the south of the region, face multiple environmental and social challenges including heatwaves, flash floods, health effects of air pollution, saltwater intrusion, and an inadequacy of energy and water resources. Although the three urban case studies are more clustered in the east and south of the region (Fig. 1.3) than the more geographically diverse rural and coastal case studies (Figs. 1.4 and 1.5), they cover diverse European and non-European socio-economic contexts and environmental vulnerabilities and have provided an excellent opportunity for using an integrated and multi-scale (temporal and spatial) approach. In the spatial dimension, research extends from the inner-city boundary to the surrounding mountains, forested areas or coast. In the temporal dimension, research extends from the observation time period to future time periods using scenarios generated from a range of climate models. In addition, a multi-sector approach to climate change impacts has been adopted which ranges from direct climate impacts on natural ecosystems to indirect impacts operating through cascading climate-socioeconomic linkages.

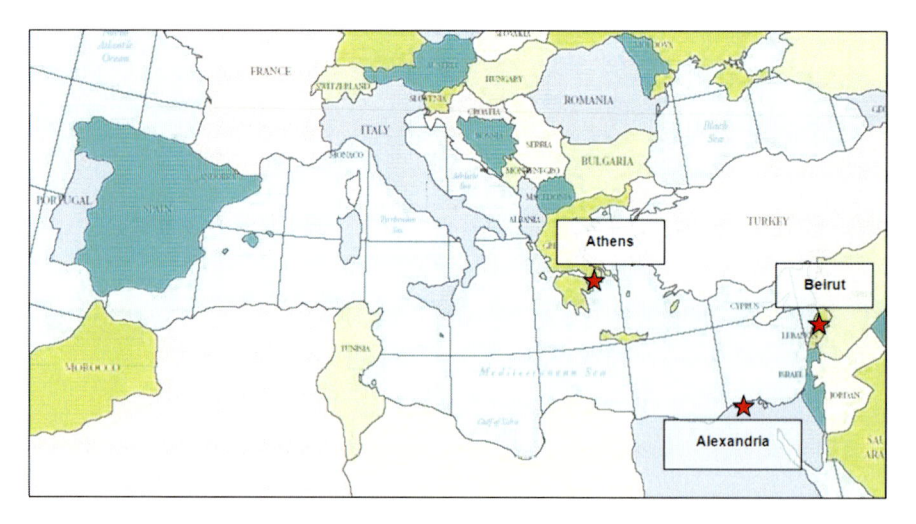

Fig. 1.3 Location map of the three urban case studies: Athens (Greece), Beirut (Lebanon) and Alexandria (Egypt)

1.4.2.1 Athens (Greece)

Christos Giannakopoulos; Dimitra Founda; Maria Hatzaki; Effie Kostopoulou; Basil Psiloglou; and Kostas Varotsos (National Observatory of Athens)

Athens is a large, densely populated metropolitan city of around four million inhabitants, with few green open spaces. The city is surrounded by mountains favoring the formation of air pollution during anticyclonic conditions. A dense city layout with narrow streets, limited green space, increased motor-vehicle traffic and long-standing air pollution problems contribute to the heat of the city in the summer which is prone to heatwaves. In addition, flash floods occur in the densely built urban areas of Athens during heavy precipitation events. The combined hazards of temperature extremes and air pollution have adverse impacts on: human health (through heat stress and respiratory disease), tourism (as heat stress, air pollution incidents and wild fires reduce destination attractiveness), energy (as heat extremes increase the demand for air conditioning) and peri-urban forest fires (as higher temperatures and protracted drought increase the fire risk) (Giannakopoulos et al. 2011).

1.4.2.2 Beirut (Lebanon)

Mutasem El-Fadel (American University of Beirut)

Greater Beirut is the largest metropolitan area in Lebanon with around two million inhabitants, representing nearly half of the country's population. It is a densely populated urban area renowned for problems of water shortage (Yamout and El-Fadel

2005). In addition to pressures of high population growth, water demand has been rising in response to tourism and industrial development. Consequently, Beirut experiences an acute water deficit leading to intermittent supply in most areas, coupled with a lack of piped water supply to a large number of socially deprived districts. During periods of water shortage, water is abstracted from private wells or purchased from unregulated vendors increasing the risk of microbial contamination. Vulnerability (and particularly health vulnerability) to climate variability and change is heightened by poverty, limited access to clean water and sanitation, high population growth rates, saline intrusion of freshwater resources, intense industrialization and a lack of rigorous environmental regulation.

1.4.2.3 Alexandria (Egypt)

Mohamed El-Raey (University of Alexandria); Hesham El-Askary (Chapman University; University of Alexandria); Mamdouh El Hattab (Minufiya University)

Alexandria is the largest city in the West Nile Delta with a population of over 3.5 million. The Greater Alexandria Area is home to more than 60 % of Egypt's industry and suffers from severe environmental pollution (due to inadequate management of solid and hazardous waste) and a lack of strong environmental monitoring and regulation infrastructure. Its deltaic location makes the city vulnerable to sea-level rise and salt-water intrusion. Aspects of the urban environment at greatest risk to climate variability and change are: residential areas, energy production facilities and industry located in the flood zone; freshwater resources (at risk from saline intrusion); and human health (at risk from heat stress and poor air quality). The integrated assessment focuses on the direct and indirect impacts of sea-level rise, shortages in water resources, heat extremes, air pollution and dust storms. The risk of sudden inundation of inhabited areas located below sea level, should for example the Mohamed Ali Sea Wall fail, is also considered.

1.4.3 Rural

Rural areas throughout the Mediterranean region sustain a multitude of functions other than merely agricultural production. They act as areas for recreation, and contribute to the region's cultural identity, tourism, and the preservation of natural resources. The fragile balance of these complex systems is threatened by environmental and social pressures, rendering them particularly vulnerable to climate change. The assessment of climate change impacts in rural areas is not straightforward since they encompass a wide range of human activity sectors (such as tourism, agriculture and forestry) in combination with processes and systems of the natural environment. The four rural case-study regions (Fig. 1.4) are characterized by a range of vulnerability features and climate-sensitive issues that make them an ideal focus for cross-sector study and integrated assessment.

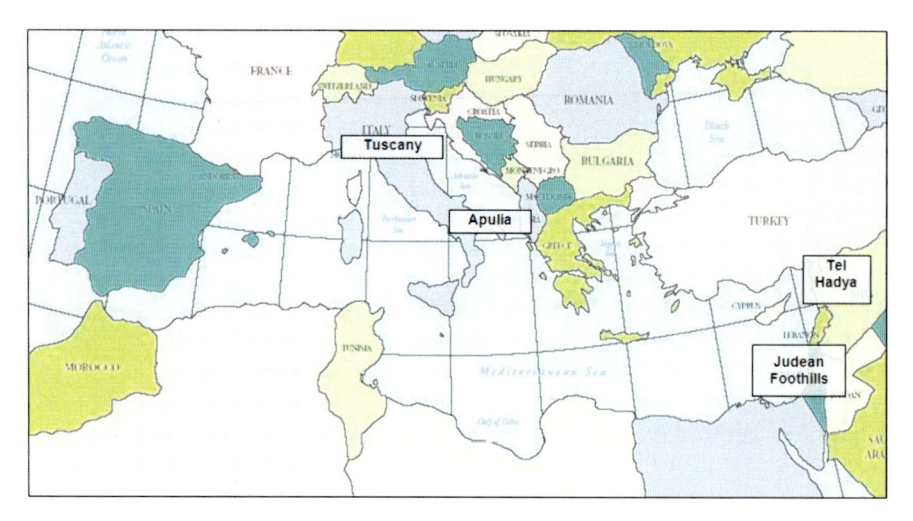

Fig. 1.4 Location of the four rural case studies: Tuscany and Apulia (Italy), Tel Hadya (Syria), and the Judean Foothills (Israel)

1.4.3.1 Tuscany, Italy

Marco Bindi; Camilla Dibari (University of Florence); Roberto Ferrise (CNR-IBIMET, Florence)

Rural areas of Tuscany are predominantly a focus of agricultural and tourist activities. Grapevine, olives and wheat are three of the main agricultural crops grown in Tuscany. The former is very important for the economic role that wine producers play at a national and international level; whilst the latter is important not only for the income that it generates but also for its contribution to landscape. Wheat fields characterize the typical Tuscan landscape, especially in the '*Maremma*' areas located in the south west of the region, and are appreciated all over the world. However, wheat (usually a rain-fed crop in Tuscany) is particularly sensitive to climate variability and change. Tuscan land use has changed notably over the last century, as large-scale abandonment of rural areas has occurred and crop fields and grazing lands have been replaced by open unmanaged forests, shrub or woodlands. Additionally in the last 15 years, rural tourism has increased substantially in Tuscany so that it now represents the main income for many farmers. Cross-sectorally, water availability represents one of the main constraints in Tuscany, with individual sectors (such as agriculture, tourism, and industry) relying on the same limited resource.

The regional administration of Tuscany has a long-standing interest in understanding climate trends and future climate impacts on key socio-economic and environmental sectors of the region. Accordingly, a variety of regional projects have been conducted within this research frame, such as the projects META and ACQUA. CLARINO (http://www.dipsa.unifi.it/CMpro-v-p-167.html) is the most extensive of these regional projects financed by the administration of Tuscany (CNR-IBIMET and DIPSA, University of Florence, are involved in this project). CLARINO evaluates the hazards and impacts of present and future climate for several sectors of the regional economy (e.g., agriculture, forestry, and livestock) within an integrated framework that also assesses the economic costs that could be incurred by the regional community to compensate for the potential consequences of climate change.

1.4.3.2 Apulia, Italy

Piero Lionello; Letizia Congedi; Cosimo Pino; Marco Reale; and Annalisa Tanzarella (University of Salento)

A gentle topography with high population density has led to an intensification of agricultural activities in Apulia (the south-eastern '*heel*' of Italy). The successive replacement of existing natural vegetation with selected crops and pasture has resulted in a loss of biodiversity. A parallel increase in irrigation crops has extended cultivation from vegetables to trees, with intensive irrigation being used to grow table grapes and citrus fruits, and emergency irrigation being used to grow olive crops by an unregulated process which has little regard for the long-term sustainability of water resources. Tourism in the Apulia region is a growing economic resource, whose future development could be constrained by a decline in water resources, summer heat extremes, and by intrusive coastal protection strategies. An understanding of water resource variability and decline in the context of human society and ecosystem adaptation is of paramount importance to the region. In parallel, temperature changes could affect crop growth, tourism and energy consumption. Knowledge of the interplay between environmental and social factors in the Apulia case-study is critical to understanding future impacts of climate change and in identifying potentially effective and regionally targeted adaptation options.

1.4.3.3 Tel Hadya, Syria

Theib Oweis; Lawand Hussein; Mohammed Karrou; and Rolf Sommer (International Center for Agricultural Research in the Dry Areas – ICARDA)

The Tel Hadya case-study (in Aleppo Province, north-western Syria) is located in the dry region of West Asia where high temperatures are experienced during the maturation stage of crop growth, and drought is frequent and can occur at any time during the crop season. The economy of the Tel Hadya region relies heavily on rain-fed agricultural systems with wheat production playing a particularly essential role in the livelihood of the rural poor. The rural communities of the area are therefore particularly vulnerable to drought, and rainfall is a crucial limiting factor. To boost and stabilize wheat yield, many farmers have dug wells to abstract groundwater for conventional irrigation but by doing so the water table has been lowered and this practice is not sustainable in the long term. This case study evaluates the effectiveness of supplemental irrigation for wheat production in Tel Hadya, as an alternative agricultural practice to conventional irrigation. The use of irrigation water at critical stages of crop growth can improve water productivity and reduce the damaging effects of drought and heat stress on grain yield (Oweis and Hachu 2006). An understanding of the long-term trends and variability in rainfall during the growing season can help farmers manage available water resources and policy makers develop water allocation plans, and improved drought adaptation and preparedness plans.

1.4.3.4 Judean Foothills, Israel

José Grünzweig and Dina Kanas (the Hebrew University of Jerusalem)

This region is located in the southern-central part of Israel and occupies the transition zone between the northern sub-humid and southern semi-arid Mediterranean climate zones. A steep moisture gradient together with wide topographical variation has created high biodiversity and considerable ecosystem variability. Land use ranges from nature reserves to forest plantations and intensively managed agriculture with small rural settlements. Wheat (largely rain fed) is the main field crop in Israel in terms of area covered, and is an important source of income for the rural economy. Pine forests provide a wide-variety of ecosystem functions including biological conservation, wood production, and areas of recreation. Tourism, a developing sector in the area, relies on the rich historical, cultural and natural heritage of the Judean Foothills. Major sources of biogeophysical vulnerability are reduced agricultural yields, a decline in tree growth in afforested systems, lower biodiversity in afforested and natural land, and an increased risk of pest and diseases. Issues of social vulnerability include rapid population growth, a heightened competition for scarce water resources, and a downturn in tourism related to heat extremes and landscape deterioration. The case study explores the risks of climate variability and change (particularly heat stress and reduced water availability) to the rural economy and considers carbon sequestration in forestry as a mitigation strategy.

1.4.4 Coastal

Coastal zones have one of the highest population densities in the Mediterranean region, support an unparalleled concentration of uses, infrastructure, and productivity, and provide a vital link between terrestrial and aquatic ecosystems. Even under present climate conditions, the four coastal case studies (Fig. 1.5) experience a high frequency of hazards since they are naturally much more dynamic than inland waters or open sea, and are in conflict with the 'rigid' civil engineering structures built within coastal zones (Sánchez-Arcilla et al. 2010). Consequently, coastal locations constitute an excellent basis for evaluation of cross-sector impacts of climate variability and change.

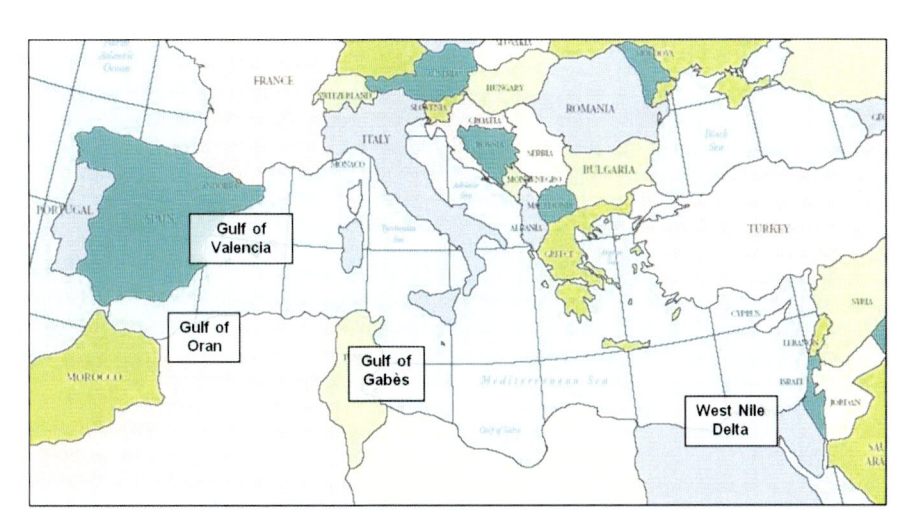

Fig. 1.5 Location of the four coastal case studies: the Gulf of Valencia (Spain), the Gulf of Oran (Algeria), the Gulf of Gabès (Tunisia), and the West Nile Delta (Egypt)

1.4.4.1 Gulf of Valencia – Catalan Coast, Spain

Agustín Sánchez-Arcilla; César Mösso Aranda; and Joan Pau Sierra (Universitat Politécnica de Catalunya)

Within the Gulf of Valencia – Catalan coast, the Ebro Delta and Cullera Bay are examples of coastal areas vulnerable to change in environmental and climate systems. The Ebro Delta is a low-lying coastal area protecting a deltaic plain of high environmental value and intense agricultural activity. The Ebro River has experienced an appreciable reduction in flow and sediment load, leading to increasing rates of coastal erosion, subsidence of the Ebro Delta and a decline in near shore water quality. Cullera Bay receives high nutrient loads from the River Júcar and a marine outfall, and Cullera Cape exerts a barrier effect on currents, propagating problems of coastal pollution. The main climate hazards for the case study are air and sea warming and storm surges, in addition potential hazards are generated by wave fields of a differing direction to prevailing conditions, for which the coast is out of equilibrium and significant shoreline changes can ensue. Key issues include coastal erosion, flooding and land loss, marine water quality, saline intrusion, and fisheries production. Flood vulnerability in the case-study region has risen as the coastal population and intensification of land use has increased, and the flood vulnerable area has expanded. High insurance losses have incurred especially following the storms of 2001, 2002 and 2008, and there has been a decline in the performance of many coastal protection structures. Some bays and harbors are experiencing a significant decrease in water quality, and there has been a commensurate decline in fisheries in the Ebro Delta zone. This case-study explores associated cross-sector coastal impacts driven by changes in atmospheric and marine physical parameters with feedbacks at multiple scales.

1.4.4.2 Gulf of Oran, Algeria

Mohamed Senouci; Salah Sahabi Abed; and Mohamed Zoheir Taleb (Association de Recherche sur le Climat et l'Environnement)

The west coast of Algeria acts as a buffer zone with Atlantic currents and oceanographic dynamics determining the spatio-temporal distribution of temperature and salinity. The littoral area is exposed to many ecological disturbances including floods and sea swell, drought and heat waves. The region has experienced periods of extreme warmth in recent years which can accelerate eutrophication, further degrade water quality, and alter biological patterns and biodiversity. Key issues of relevance to the Gulf of Oran case study are biological invasion (by species such as jelly fish and green algae), food web changes, and the collapse of fish stocks. Prey species distribution and abundance could be severely altered by global warming, increasing competition among fishermen to the detriment of the wider marine ecosystem. Social pressures are associated with harbor activities, scarcity of urban waste water treatment, and concentrations of river nutrient loads and marine pollutants. The altered sea-water composition has likely deleterious effects not only on the organisms inhabiting these areas, but also on human health, and adversely contributes to the overall climate vulnerability of the coastal ecosystem. This case study also examines the role of nature reserves which act as both an ecological buffer and an experimental laboratory (and observatory) to assess species and habitat response to climate variability and change.

1.4.4.3 Gulf of Gabès, Tunisia

Ali Harzallah; Mohamed Nejmeddine Bradai; Abdallah Hattour; and Skander Ben Salem (Institut National des Sciences et Technologies de la Mer)

The Gulf of Gabès is one of the most biologically productive zones in the eastern Mediterranean and is one of the main centers for fishing activity in Tunisia. A key feature of the Gulf of Gabès is the shallowness of the basin which means that near-shore waters are very sensitive to atmospheric conditions. Changes in the physical characteristics of marine water, and in particular temperature, may have negative consequences for the fauna and flora of the gulf. The warmer waters pose a potential risk to human health, tourism and fisheries (e.g., through jellyfish and phytoplankton blooms). They also create conditions favorable to invasive species, and reduced biodiversity with potential negative consequences for fishing activity. Biologically, some species may not support an excessive temperature increase. During heat waves the most vulnerable species (particularly the sedentary ones) may die, for example, following toxic phytoplankton blooms. Jelly fish blooms may also have negative effects on fish stocks. In addition, the Gulf has a heightened vulnerability to sea-level rise due to its low elevation, particularly the low-lying islands of Kerkean, Kneis and Djerba (a key tourist destination). Severe erosion of the littoral zone has already been observed in the case-study area and this could accelerate with sea-level rise. In addition, exceptional climate-marine events such as severe wave storms are likely to cause abrupt changes in coastline.

1.4.4.4 West Nile Delta, Egypt

Mohamed El-Raey (University of Alexandria); Hesham El-Askary (Chapman University; University of Alexandria); Mamdouh El Hattab (Minufiya University)

Population, agriculture and economic activity in Egypt are constrained to a narrow fertile strip of land either side of the Nile and the deltaic coast that includes the West Nile Delta. The case-study area is extremely sensitive to climate change through its impacts on sea-level rise and water availability (El-Askary et al. 2009; Elsharkawy et al. 2009). Climate vulnerability is exacerbated through serious environmental stresses that include natural land subsidence, rapid population growth and urbanization, insufficient institutional capacity, and widespread pollution. A lack of environmental awareness acts as a barrier to sustainable development in the region. Since the building of the Aswan High Dam in 1964 there has been a rapid reduction in the amount of sediment accreted, leading to significant and rapid changes along the northern shoreline. The most vulnerable areas in the West Nile Delta are the low-lying districts of Alexandria (an urban case-study location) and Behaira governorates, and the coastline between Abu Qir Bay and the Rosetta promontory. These areas are highly vulnerable to the combined hazards of sea-level rise and heat extremes, which increase the risk of saline intrusion, water shortages for agriculture and industry, lower crop yields, flooding and inundation of prime agricultural land and industrial facilities, and insurance losses due to storm damage and coastal flooding.

One aspect common to all of the CIRCE case studies and one of their selection criteria (Sect. 1.1) was access to regional stakeholders including decision and policy makers. The next chapter provides details of the generic types of stakeholders that were involved in each of the case studies, the process by which they were involved in the project, and the key areas of contribution to the case-study assessments.

References

Adger WN (2006) Vulnerability. Glob Environ Chang 16(3):268–281. doi:10.1016/j.gloenvcha.2006.02.006

Allen HD (2003) Response of past and present Mediterranean ecosystems to environmental change. Prog Phys Geogr 27(3):359–377. doi:10.1191/0309133303pp387ra

Brooks N (2003) Vulnerability, risk and adaptation: a conceptual framework. Tyndall Centre working paper no. 38, September 2003, Tyndall Centre for Climate Change Research University of East Anglia, Norwich, UK, 20 pp. http://www.tyndall.ac.uk/sites/default/files/wp38.pdf

Brooks N, Adger WN, Kelly MP (2005) The determinants of vulnerability and adaptive capacity at the national level and the implications for adaptation. Glob Environ Chang 15:151–163. doi:10.1016/j.gloenvcha.2004.12.006

Carter TR, Ones RN, Lu JX, Bhadwal S, Conde C, Mearns LO, O'Neill BC, Rounsevell MDA, Zurek MB (2007) New assessment methods and the characterisation of future conditions. In: Parry ML, Canziani OF, Palutikof JP, van der Linden PJ, Hanson CE (eds) Climate change 2007: impacts, adaptation and vulnerability. Contribution of working group II to the fourth assessment report of the intergovernmental panel on climate change. Cambridge University Press, Cambridge, pp 133–171

Conde C, Lonsdale K (2005) Engaging stakeholders in the adaptation process. In: Lim B, Spanger-Siegfried E, Burton I, Malone E, Huq S (eds) Adaptation policy frameworks for climate change: developing strategies, policies and measures. Cambridge University Press, Cambridge/New York, pp 47–66

Dessai S, Hulme M, Lempert R, Pielke R Jr (2009) Climate prediction: a limit to adaptation? In: Adger WN, Lorenzoni I, O'Brien KL (eds) Adapting to climate change: thresholds, values, governance. Cambridge University Press, Cambridge, pp 64–78

Downing TE, Patwardhan A (2005) Assessing vulnerability for climate adaptation. In: Lim B, Spanger-Siegfried E, Burton I, Malone E, Huq S (eds) Adaptation policy framework for climate change: developing strategies, policies and measures. Cambridge University Press, Cambridge/New York, pp 67–90

El-Askary H, Hattab M, El-Raey M, Kafatos M (2009) Impact of sea level rise on the low land area South East of Alexandria, Egypt. Paper presented at the American Geophysical Union, Fall Meeting, San Francisco, 14–19 December 2009, abstract #NH11A-1097

Elsharkawy H, Rashed H, Rached I (2009) The impacts of SLR on Egypt. Paper presented at the Low Carbon Cities, 45th ISOCARP international congress, Porto, Portugal, 18–22 October 2009, 11 pp

Füssel H-M, Klein RJT (2006) Climate change vulnerability assessments: an evolution of conceptual thinking. Clim Chang 75(3):301–329. doi:10.1007/s10584-006-0329-3

Gambaiani DD, Mayol P, Isaac SJ, Simmonds MP (2009) Potential impacts of climate change and greenhouse gas emissions on Mediterranean marine ecosystems and cetaceans. J Mar Biol Assoc U K 89(1):179–201

Georgas DG (2003) Assessing the impacts of climate change in the Mediterranean coastal zones. In: Giupponi C, Schechter M (eds) Climate change in the Mediterranean: socio-economic perspectives of impacts, vulnerability, and adaptation. Elgar, Cheltenham/Northampton, pp 225–244

Giannakopoulos C, Bindi M, Moriondo M, LeSager P, Tin T (2005) Climate change impacts in the Mediterranean resulting from a 2°C global temperature rise. A report for WWF, the global conservation organization, Gland, Switzerland, 1 July 2005, 67 pp

Giannakopoulos C, Le Sager P, Bindi M, Moriondo M, Kostopoulou E, Goodess CM (2009) Climatic changes and associated impacts in the Mediterranean resulting from a 2°C global warming. Glob Planet Chang 68(3):209–224. doi:10.1016/j.gloplacha.2009.06.001

Giannakopoulos C, Kostopoulou E, Tziotziou K, Varotsos KV, Plitharas A (2011) An integrated assessment of climate change impacts in Greece in the near future. Reg Environ Chang 11:829–843. doi:10.1007/s10113-011-0219-8

Giupponi C, Schechter M (2003) Climate change in the Mediterranean: socio-economic perspectives of impacts, vulnerability, and adaptation. Elgar, Cheltenham/Northampton, 322 pp

IPCC (2007) In: Parry ML, Canziani OF, Palutikof JP, van der Linden PJ, Hanson CE (eds) Climate change 2007: impacts, adaptation and vulnerability. Contribution of working group II to the fourth assessment report of the intergovernmental panel on climate change. Cambridge University Press, Cambridge, 976 pp

Mellouki A, Ravishankara AR (eds) (2007) Regional climate variability and its impacts in the Mediterranean area. In: Proceedings of the NATO advanced research workshop on regional climate variability and its impacts in the Mediterranean area, Marrakech, Morocco, 1 September 2006, NATO science series IV: Earth and environmental sciences, vol 79. Springer, London, 316 pp. ISBN 1402064276

Milano M (2010) The foreseeable impacts of climate change on the water resources of four major Mediterranean catchment basins. Plan Bleu Regional Activity Centre UNEP/MAP, Sophia Antipolis, January 2010, 6 pp

Moriondo M, Good P, Durao R, Bindi M, Giannakopoulos C, Corte-Real J (2006) Potential impact of climate change on fire risk in the Mediterranean area. Clim Res 31(1):85–95. doi:10.3354/cr031085

Niemeijer D, de Groot R (2008) A conceptual framework for selecting environmental indicator sets. Ecol Indic 8(1):14–25. doi:10.1016/j.ecolind.2006.11.012

Olesen JE, Carter TR, Díaz-Ambrona CH, Fronzek S, Heidmann T, Hickler T, Holt T, Minguez MI, Morales P, Palutikof JP et al (2007) Uncertainties in projected impacts of climate change on European agriculture and terrestrial ecosystems based on scenarios from regional climate models. Clim Chang 81(Supplement 1):123–143. doi:10.1007/s10584-006-9216-1

Oweis T, Hachu A (2006) Water harvesting and supplemental irrigation for improved water productivity of dry farming systems in West Asia and North Africa. Agric Water Manag 80(1–3):57–73. doi:10.1016/j.agwat.2005.07.004

Sánchez-Arcilla A, Mösso C, Sierra JP, Mestres M, Harzallah A, Senouci M, El Raey M, El-Askary H (2010) Climatic drivers of potential hazards in Mediterranean coasts. Reg Environ Chang 11:617–636. doi:10.1007/s10113-010-0193-6

Taubenböck H, Post J, Zosseder K, Strunz G, Dech S (2008) A conceptual vulnerability and risk framework as outline to identify capabilities of remote sensing. Nat Hazards Earth Syst Sci 8:409–420

UNDP (2005) In: Lim B, Spanger-Siegfried E, Burton I, Malone E, Huq E (eds) Adaptation policy frameworks for climate change: developing strategies, policies and measures. Cambridge University Press, Cambridge/New York, 258 pp. http://www.undp.org/gef/undp-gef_publications/undp-gef_publications.html

Valls J-F, Sardá R (2009) Tourism expert perceptions for evaluating climate change impacts on the Euro-Mediterranean tourism industry. Tour Rev 64(2):41–51

Wilby RL, Troni J, Biot Y, Tedd L, Hewitson BC, Smith DM, Sutton RT (2009) A review of climate risk information for adaptation and development planning. Int J Climatol 29:1193–1215. doi:10.1002/joc.1839

Yamout G, El-Fadel M (2005) An optimization approach for multi-sectoral water supply management in the Greater Beirut Area. Water Resour Manag 19(6):791–812. doi:10.1007/s11269-005-3280-6

Chapter 2
Stakeholders

Maureen D. Agnew, Clare M. Goodess, Debbie Hemming,
Christos Giannakopoulos, Marco Bindi, Camilla Dibari, Hesham El-Askary,
Mamdouh El-Hattab, Mohamed El-Raey, Roberto Ferrise, Ali Harzallah,
Maria Hatzaki, Effie Kostopoulou, Piero Lionello, Salah Sahabi Abed,
Agustín Sánchez-Arcilla, Mohamed Senouci, Rolf Sommer,
Mohamed Zoheir Taleb, and Annalisa Tanzarella

Abstract The active participation of the stakeholder community is viewed as a crucial consideration in the regional assessment of climate change in the Mediterranean. Each of the CIRCE integrating case studies has benefitted from the real-life experience and expertise of local and regional stakeholders. The level of involvement has included stakeholder dialogue on an informal basis throughout the project, culminating in a series of more formal regional stakeholder workshops in the final year of the project. For each case study, stakeholders were chosen from the arena for which decisions are made at the regional-to-local level, and comprise local and regional decision and policy makers within government departments, agencies,

---

M.D. Agnew (✉) • C.M. Goodess
Climatic Research Unit, School of Environmental Sciences,
University of East Anglia, Norwich, UK
e-mail: m.agnew@uea.ac.uk

D. Hemming
Climate Impacts Analysis Team, Met Office Hadley Centre, Exeter, Devon, UK

C. Giannakopoulos • M. Hatzaki • E. Kostopoulou
Institute for Environmental Research and Sustainable Development,
National Observatory of Athens, Athens, Greece

M. Bindi • R. Ferrise
Department of Plant, Soil and Environmental Science (DIPSA),
University of Florence, Florence, Italy

Institute for Biometeorology, National Research Council of Italy,
(CNR-IBIMET), Florence, Italy

C. Dibari
Department of Plant, Soil and Environmental Science (DIPSA),
University of Florence, Florence, Italy

and public/private sector entities. Regional stakeholders have contributed to the case-study assessments in six key areas: the conceptual framework and indicator selection, data and knowledge, identification of climate impact thresholds, risk assessment and management, strategies for adaptation, and guidance for policy. However, impediments to stakeholder participation have also been highlighted and include lack of motivation and resources commitment, stakeholder fatigue, and a need for targeted information. The CIRCE experience suggests that a more participatory approach, which involves sharing of 'best' practice and accessible targeted information, will be fundamental to successful adaptation planning in the Mediterranean region. This experience has also been used to develop a good practice checklist for facilitating stakeholder involvement in research projects.

Keywords Climate change • Mediterranean • Stakeholders • Policy • Adaptation

---

H. El-Askary
School of Earth and Environmental Sciences, Schmid College of Science and Technology,
Chapman University, Orange, CA, USA

Department of Environmental Sciences, Faculty of Science,
Alexandria University, Alexandria, Egypt

M. El-Hattab
Environmental Studies and Research Institute (ESRI-SADAT),
Minufiya University, Sadat City, Egypt

M. El-Raey
Department of Environment Studies, University of Alexandria,
Alexandria, Egypt

A. Harzallah
Institut National des Sciences et Technologies de la Mer – INSTM,
Salammbô, Tunisia

P. Lionello
Centro Euro-Mediterraneo per i Cambiamenti Climatici,
CMCC, Lecce, Italy

Department of Material Science, University of Salento, Lecce, Italy

S. Sahabi Abed • M. Senouci • M.Z. Taleb
Association Recherche Climat et Environnement, ARCE, Oran, Algeria

A. Sánchez-Arcilla
Laboratori d'Enginyeria Marìtima (LIM/UPC), Universitat Politécnica de Catalunya,
Barcelona, Spain

R. Sommer
International Center for Agricultural Research in the Dry Areas,
ICARDA, Aleppo, Syria

A. Tanzarella
Department of Material Science, University of Salento, Lecce, Italy

2.1 Introduction

In scientific research, the principal contribution of stakeholders is in the provision of individual and organizational knowledge and expertise. However, stakeholder dialogue should not be viewed as a simple linear unidirectional exchange of information, but as a two-way process of communication operating at multiple levels and scales (Kasperson 2005; Vogel et al. 2007). Within the scientific discourse of the CIRCE project a methodology was developed to take into account stakeholders needs, priorities and questions. Indeed, the active inclusion of the stakeholder community has been viewed as a prime consideration in producing the regional assessment for climate change in the Mediterranean. Each of the integrating case studies has capitalized on the real-life experience and expertise of local and regional stakeholders.

Strategies for stakeholder dialogue have taken a variety of forms in previous and current climate impact and vulnerability assessments. Some have placed an emphasis on stakeholder consultation towards the end of the project (e.g., Hanson et al. 2006), or in the provision of knowledge transfer through end-of-project non technical reports (e.g., the EU project CLIMB (http://www.climb-fp7.eu) which considers climate-induced changes on the hydrology of Mediterranean basins, in the context of reducing uncertainty and quantifying risk). During the course of the EU ENSEMBLES project, local (Athens, September 2008) and international (Venice, December 2008) stakeholder workshops were organized when the project was over halfway complete. Some EU projects have placed a strong emphasis on stakeholder participation. For example, the EU project FIELD_AC (http://lim050.upc.es/field_ac) aims to provide high-resolution forecasts for waves, currents and wind in the coastal zone, and is working in close cooperation with targeted private/public end users. One of the chosen case studies, the Catalan Coast, is the same location as a coastal CIRCE case-study site. The CIRCE case studies have aimed at a participatory level of involvement and have included stakeholder dialogue on an informal basis throughout the project, culminating in a series of more formal regional stakeholder workshops in the final year of the project.

The category of stakeholders participating and contributing to the CIRCE case studies has been relatively focused and can be identified by the arena in which decisions related to climate change are made at the regional and local level. In particular, the focus of the CIRCE case studies has been on a practitioner level rather than on a higher strategic policy level. The latter was the focus of the CIRCE-wide stakeholder workshop (Paris, 2007), designed as a high-level consultation among a small group of selected participants representing the agriculture, energy and tourism sectors. Discussion centered on the two themes of ecosystem protection, and urban and regional planning, to determine on a general level 'what was really at stake' in the context of climate change in the Mediterranean region. In contrast, the CIRCE case studies were focused on the detail of climate change, impacts, vulnerability and adaptation at a regional and local scale. The corresponding stakeholder community comprised local and regional decision and policy makers within government departments, agencies, and public/private sector entities (Table 2.1).

Table 2.1 Categories of stakeholders relevant to the generic integrating case studies

Case study	Sector	Generic stakeholders
Urban	Urban planning	Planners/developers/councils
	Health	Local or regional government/hospital managers/consultants
	Industry	Managers/investors
	Energy	Power corporations
		Local or regional government
	Water	Local or regional government
	Environment	Local or regional government/agency
	Tourism	Local or regional government
	Civil protection	Local or regional government/agency
	Society	General public
	Research	Research councils/academic institutions/private consultants
Rural	Environment/conservation	Local or regional government/agencies/international governmental organizations (UNESCO)/academic institutes
	Agriculture	Farmer consortiums/farmer union/producers/vintners/local or regional government
	Forestry	Local or regional government/agency/charitable conservation organizations and societies
	Water	Local or regional government
	Energy	Local or regional government
	Health	Local or regional government
	Planning	Local or regional government
	Tourism	Local or regional government
	Economy	Financial organizations/regional agencies
	Insurance	Company managers/insurance associations
	Research	Academic institutions
	Society	Rural communities
Coastal	Fisheries	Fishermen/guilds/authorities/producer unions/local or regional government
	Environment	Local or regional government/agency
	Rivers	Authorities/agencies
	Coasts/harbors	Authorities/agencies
	Waste	Authorities/agencies
	Agriculture	Farmers/academic institutions
	Planning and development	Local or regional government/planners/developers
	Economy	Local or regional government
	Tourism	Local or regional government
	Research	Academic institutions

2.2 Level of Stakeholder Involvement

There are a wide variety of approaches and levels at which stakeholder involvement may take place, from the lowest level of passive engagement at which stakeholders solely supply information, to the highest level at which stakeholders actively design

```
┌─────────────────────────────────────────────┐
│ Self Mobilisation: stakeholders             │──────────┐
│ initiate the assessment and control         │          │
│ its process, contracting skilled            │          │
│ assistance to participate where             │          │
│ needed                                      │          │
├─────────────────────────────────────────────┤          │
│ Partnership: stakeholders act as equal      │──────────┤
│ partners in process, contributing resources │          │
│ and taking ownership of the outcomes        │          │
├─────────────────────────────────────────────┤          │
│ Participation: stakeholders collaborate in process driven│
│ by a third party and are dependent on outside resources │
│ but can act to ensure process and results are relevant  │
├─────────────────────────────────────────────┤          │
│ Consultation: stakeholders are consulted by researchers to provide│
│ information which is then used to tailor the assessment process.  │
├─────────────────────────────────────────────┤          │
│ Information: stakeholders are involved in interviews or questionnaires but have no│
│ influence on the process                                          │
└─────────────────────────────────────────────┘
```

Catalysing Change: Stakeholders use their experience to influence other stakeholders and subsequent assessments

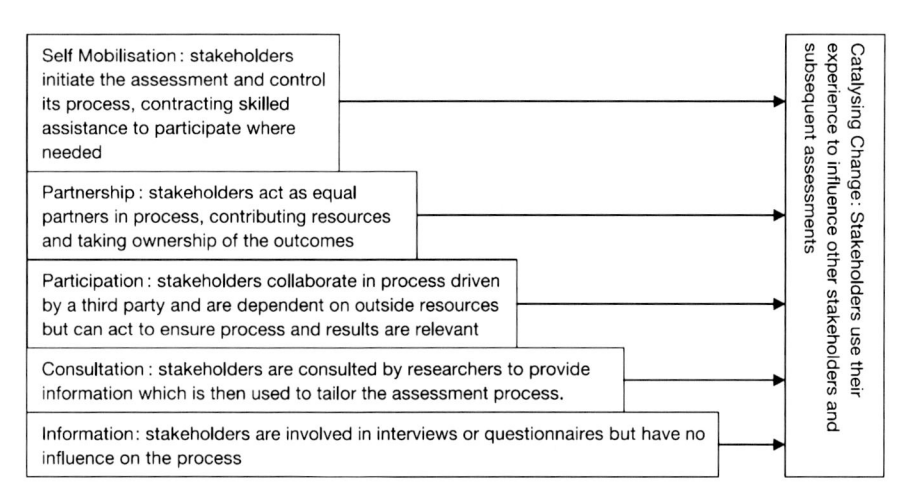

Fig. 2.1 Ladder of stakeholder participation (From Carter et al. 2007, based on Pretty et al. 1995, and Conde and Lonsdale 2005)

and initiate the process (Fig. 2.1). At an intermediary level, stakeholders may be consulted or act as equal partners in the research program, such as in the UNECE Water Convention (http://www.unece.org/env/water/) which in 1999 established a framework to encourage trans-boundary water cooperation and security across Europe, and required all main stakeholders to act as partners. At the highest level of self mobilization, stakeholders take the initiative. The EU project CLIMSAVE (http://www.climsave.eu/climsave/index.html) is developing and implementing an integrated web-based methodology for stakeholder-led climate change impact and vulnerability assessments for landscape change across Europe. Higher levels of participation consider linkages and networks of stakeholders, government agents, communities and organizations at multiple levels and sectors (Moser 2008, 2009). Multi-level stakeholder networks are proposed, for example, as a means of determining effective modes of governance for renewable energy in responding to the dual challenges of climate change and development (Benecke 2010).

The CIRCE case studies have aimed at the level of stakeholder participation in which, 'stakeholders collaborate in process driven by a third party and are dependent on outside resources but can act to ensure process and results are relevant' (Fig. 2.1). Stakeholder dialogue has been an ongoing component of the research process in the Mediterranean case studies, and has involved several cycles or phases. Initiating contacts with relevant stakeholders and agreeing participation within the project represents a preliminary phase in the stakeholder participatory process. Contributions from subsequent phases are outlined in Sect. 2.4. Initial cycles aimed to identify key research issues in the case-study regions through informal discussion; these progressed to information review cycles, and more formal consultation on impacts, adaptation and mitigation, and policy recommendations in a workshop forum. Across the case studies there has been a degree of variation in the level of involvement, e.g., from a general discussion of impacts and adaptation to climate change in the region, to direct involvement in indicator selection, exchange of ideas, the provision of data, and testing and evaluating measures of adaptation (using a variety of tools and strategies). Dissemination strategies and stakeholder resources included a series of briefing notes (providing contextual information and key research issues for each of the case studies), a set of case-study information sheets (Sect. 3.2.5), and a dedicated CIRCE case-studies web site (http://www.cru.uea.ac.uk/projects/circe/index.html) from which the briefing notes and information sheets can be downloaded and further stakeholder relevant information is accessible on vulnerability, impacts and adaptation.

2.3 Objectives of Stakeholder Involvement

The CIRCE case-study approach provided valuable opportunity to identify and explore specific regional and city level climate change issues and adaptation strategies using both 'top-down' (based on literature review and the modeling expertise of the scientific research teams) and 'bottom-up' (based on the real-life experiences and expertise of regional stakeholders) approaches. Stakeholder involvement during the early project stages helped focus attention on key climate hazards and sensitivities, while involvement at later stages allowed discussion of relevant social and biogeophysical trends and impacts, future climate projections, and adaptation strategies.

Stakeholder involvement was enhanced by holding regional workshops and producing focused pre-workshop background material, in some cases in local languages. These workshops largely took place in the final stages of the project, and provided an arena for exchanging information on past and present trends in climate hazards at the level of cities or regions; and discussion of integrated impacts and vulnerabilities of climate variability and change. Informal stakeholder consultation and meetings were also conducted in association with other meetings, with representatives from government institutions, and colloquia at national and internal levels. The Oran case-study team, for example, held a series of meetings and workshops involving government institutions such as the *Regional Direction of Environment and Planning* and the *Regional Direction of Tourism* in Oran 6 months prior to the larger and more formal CIRCE regional workshop. These meetings facilitated an evaluation of climate impacts and local policy. In addition, dialogue with regional stakeholders and decision makers has been supported using poster sessions presenting preliminary results from the case study, such as at the *2nd International Colloquium on Biodiversity and Coastal Ecosystems* hosted by the University of Oran and other national and international partners (http://lrse-bel2.webnode.fr).

A number of tools and techniques to facilitate stakeholder interaction were developed during the course of the project. The societal dynamics project research line developed several decision-making tools including elicitation and visualization tools following a Bayesian approach. In addition, the induced policies research line developed an adaptation options and screening tool (Part II, Chap. 12) within the framework of an interactive dynamic website, www.wikiADAPT.org. The website has particular application to stakeholder interaction and encourages the sharing of information and participatory techniques with an underpinning belief that adaptation

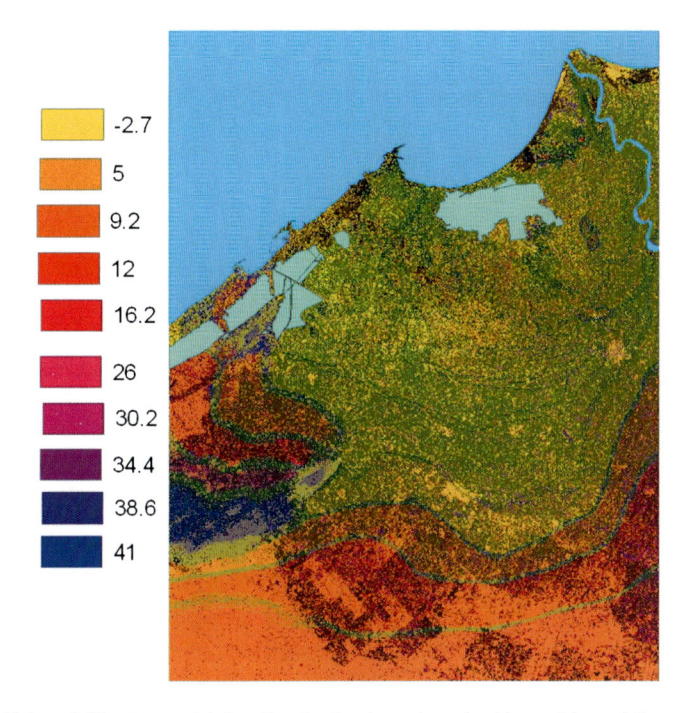

Fig. 2.2 Vulnerability to coastal flooding by land-use type in Alexandria and the surrounding region, West Nile Delta, Egypt. High spatial resolution digital elevation map in meters (see legend) was used to create contours of elevation. This elevation layer was overlain with a land cover map within a Geographical Information System to identify zones at high risk from coastal flooding

involves a socio-institutional process of learning and change that requires innovative means of support. Appropriate application of these tools depends implicitly on the willingness and cooperation of the research teams, and the adaptability of the tools to the case-study situation. The Alexandria and West Nile Delta case-study teams used the weADAPT platform in stakeholder engagement to help build capacity for climate change adaptation. Complexity of use, software costs, time constraints, and language barriers prevented these tools from being fully utilized in the stakeholder workshops. However, several ideas from the online questionnaire tool were incorporated in the pre-workshop questionnaire designed to directly extract experts' opinion, and wikiADAPT includes links to CIRCE case-studies information.

A number of case-study specific tools and techniques were tested within the regional stakeholder workshops. These included visual tools such as case-study Geographic Information Systems (GIS; used in the Gulf of Gabès and West Nile Delta coastal case studies, and the Alexandria urban case study) featuring layers of relevant spatial information, and perception questionnaires (e.g., in the Gulf of Oran coastal case study and Tuscany rural case study) to assess stakeholder views on the relative risks posed by future climate variability and change. In the Alexandria and West Nile Delta studies, six land cover types were combined with high spatial resolution elevation data within a GIS environment to provide stakeholders with a single map illustrating coastal flooding vulnerability by land cover (Fig. 2.2). The proportion of land cover class in each zone of elevation was subsequently calculated and used in stakeholder discussion of policy action. The Valencia-Catalan coastal case study has highlighted the importance of audio-visual resources in efficiently communicating climate information to stakeholders and the general public.

Focus groups, the 'world café' conversational processes, and plenary sessions were employed during the local workshops to consider the specific issues related to climate change for each case study, understand management of current hazards of climate variability and change, and identify and evaluate targeted adaptation-mitigation measures and the barriers (present or potential) to their successful implementation. The adaptation measures considered encompassed planned and anticipatory adaptation to climate change projections and responsive or autonomous adaptation to experienced weather events.

The World Café (TWC: http://www.theworldcafe.com/) is a participatory method by which an informal café is established for participants to explore some particular issue in small-table discussion groups. Conversation continues for multiple rounds of 20–30 min. Participants begin discussing an issue around one table and at regular intervals they move to a new table. Two participants (the table host and the facilitator) remain and summarize the previous conversation to the newly arrived participants. By circulating participants around the room the conversations at each table are cross-fertilized with ideas from other tables. The collective discussion insights at each table are then presented at the end of the workshop during a final plenary session. The method ensures active participation of all participants, due to the small number of participants (a maximum of 10) permitted at each working group table. A short discussion time and thematic rotation retains participant interest and concentration, and the appointment of a host and facilitator ensures that discussion focuses on the topic, that everyone is encouraged to participate, and that the results of the discussion are effectively communicated. During the CIRCE stakeholder workshop in Tuscany two world café tables were established: one on climate and agriculture (specifically vineyards and wheat) and another on tourism (tourism in general and rural tourism). This technique was found to be very successful in facilitating discussion and encouraging the participation of all stakeholders.

2.4 Stakeholder Contribution to the Case Studies

The key contribution of stakeholders to the case-study integrated assessment of climate change in the Mediterranean has been in the provision of local ground-level knowledge and expertise. Six key areas of stakeholder contribution have been recognized in the integrating case studies: the conceptual framework and indicator selection, data and knowledge, identification of climate impact thresholds, risk assessment and management, strategies for adaptation, and guidance for policy.

2.4.1 Conceptual Framework and Indicators

The defining vulnerability issues and relevant climate hazards were structured within the context of the CCIF (Chap. 1) with the aid of the expertise of the scientific and stakeholder community. Potential sets of indicators (Chap. 3) were subsequently identified, reviewed and refined in consultation with the project's thematic research groups and, where resources permitted, with regional stakeholders. The criteria for indicator selection included relevance to policy makers and stakeholders, and end-user resonance, to ensure an indicator is readily understood and interpretable by both decision-makers and scientists.

Regional stakeholder consultation helped shape case-study indicator sets, provide a focus, and identify knowledge gaps. For the Gulf of Gabès case study, previous and ongoing studies using thematic indicators (e.g., climate change impacts on the tourism sector) proved helpful for defining indicators through stakeholder discussion and participation at various meetings. For the Tuscany rural case study, regional stakeholders agreed with the climate indicators selected by the research team but they also suggested a further key climate indicator to be analyzed, the number of rainy days during the sowing season. This indicator is crucial for sowing practices, especially in respect to winter cereals. For example, if rain is overly prolonged, it may be worthwhile sowing alternative crops (e.g., sunflower), or it may be better not to sow at all and leave fields uncultivated for a year. Stakeholders' perception was that further research was needed, both for the identification of indicators suitable for the tourism sector, and for their assessment. Energy and water consumption were identified as potentially valuable indicators, since both have a large role in the income of the rural community, and are climate sensitive.

2.4.2 Data and Knowledge

Within the CIRCE project, stakeholder collaboration has been important for gaining access to data, insight or knowledge that would otherwise have been difficult to obtain and might have been left unknown. This has included undocumented local or sector knowledge of vulnerability and risks of climate hazards. For the Gulf of Gabès case study, relevant data sets (such as precipitation, sea-level height, and tourism sector statistics) were provided by several institutions and served as qualitative diagnostics. The Apulia case study received data from the regional branch of a national agency and from a statistical institute. For the Tuscany case study, discussion with stakeholder organizations provided greater insight into current climate impacts. For example, stakeholders proposed that the preliminary study of relationships between tourist fluxes (arrivals and tourist nights) and temperatures should be extended, in particular focusing on the 'shoulder seasons' (i.e., the months of April to June, and September and October) for seaside tourism, to avoid potential 'high season' biases and to improve evaluation of temperature effects. It was also suggested that the effects of religious holy days and public holidays (both in the tourist destination region and in the source country) should be included to improve detection of the climate signal. Stakeholder dialogue in the case of Alexandria proved to be particularly useful in identifying knowledge needs and gaps where further research investment should be channeled.

Engaging stakeholders in consultation and collaboration with other projects has proved a useful approach within the CIRCE case studies. The Apulia rural case study's experience was that cooperative stakeholder consultation with other projects and CIRCE research lines could provide useful regional insights on climate impacts for agriculture and water resources. For the Gulf of Valencia – Catalan coastal case study, the strategy for knowledge transfer and encouraging stakeholder involvement was to present the new 'products' generated within the project at a higher spatial resolution than previously available. CIRCE case-study projections (decadal time scale) were presented at an end user workshop, together with short-term marine forecasts from the FIELD_AC project (Sect. 2.1) to illustrate the utility of considering differing time scales.

2.4.3 Identification of Thresholds

Thresholds or limits of tolerable risk (Sect. 3.2.3) and 'tipping points' (Sects. 2.4.6 and 6.2.1) were in some cases difficult to identify from available data, or it was not known whether for example, agricultural thresholds identified through literature for other crops or geographical regions, were appropriate to the case-study region. For some case studies, stakeholder dialogue was able to help identify or clarify such thresholds. For example, during the EU ENSEMBLES project local stakeholder workshop (Athens, September 2008), the estimated value for minimal level energy demand was confirmed by participating stakeholders and compared with estimates used in other European locations (in the north and south of Europe). In the Gabès coastal case study, stakeholders identified an increase in air temperature of 2–3°C as unfavorable for recreation and tourism activities in the region.

Reviews of literature (e.g., Porter and Gawith 1999) suggest that climate thresholds can be defined for specific crops, cultivars, and developmental stages. While stakeholders in the Tuscany rural case study could not quantify specific climate thresholds they were able to provide further ground-level information on critical stages of vulnerability and climate hazards (sudden and slow onset events, see Box 1.1) for perennial (vineyard and olive) and annual (cereal) crops. For example, stakeholders perceived that due to warming-induced shortening of phenological stages in Tuscany, the risk of late frost has become a serious threat especially to vineyards and olive. For cereals, there are two key climate hazards affecting farmers' activity. The first is a delay to sowing due to prolonged rainfall; beyond February-March sowing may not be deemed worthwhile due to the corresponding loss of production. The second complex climate hazard is higher temperature coupled with reduced precipitation around anthesis (flowering) and grain-filling phases which can seriously compromise yield.

2.4.4 Risk Assessment and Management

Understanding how a system or community manages the risks of current climate hazards provides a basis for the assessment of future climate risks. Part of this process has involved recognition of the sources and levels of uncertainties involved in decision management at the local and regional level. In addition, information is required on the time horizon over which planning decisions are made. Dialogue with farmers in Tuscany has identified a greater interest in assessing short-term rather than long-term climate risk. The provision of information through specific agro-meteorological bulletins and technical training events specifically tailored to the needs of farmers has also been suggested. In contrast, there is increasing interest in long-term planning for the rural tourism sector. In Tuscany, the *Regional Observatory on Tourism* provides support to regional planning and decision-making processes that is more focused on sustainability issues, but is showing increasing interest in climate change initiatives. In the Alexandria case study, the risk of sudden earthquake damage to the Mohamed Ali sea wall was identified as an additional hazard to climate, increasing vulnerability to sea-level rise. Enforcement of the wall by sand over a long distance was selected as the most appropriate defense strategy (Sect. 6.3.1). In the Valencia coastal case study the equitable manner in which the perceived risks of climate change are managed is an important issue, since the coastal zone is subject to multiple interacting and complex processes and impacts affecting multiple agents (e.g., tourism, farming, aquaculture, and insurance industries), with potentially conflicting needs and expectations.

2.4.5 Adaptation Strategies

Stakeholders have had crucial input to the process of identifying current coping strategies, and evaluating a range of potential adaptive options. Understanding stakeholder attitudes to risk and the underpinning values of individuals and communities has highlighted both potential social barriers and facilitators to adaptation. The following example is given for adaptation strategies considered in the Tuscany case study. Key adaptation messages for this and other case studies are presented in Sect. 6.3.1.

During the Tuscany workshop, the invited stakeholders discussed and suggested potential adaptation strategies specific to the needs of the region. For example, it was suggested that the use of later crop varieties could be a feasible solution to the delayed sowing dates for winter cereal (Sect. 2.4.3). However, to fulfill the quality requirements of the cereal industry, further research on genetic improvements would be essential. For vineyards threatened by late frosts (Sect. 2.4.3), stakeholders suggested the introduction of frost-resistant rootstock as a potential solution. The current regional development model in Tuscany is based on such concepts as quality, high living standards (with low environmental impact), a very specific cultural lifestyle (an integral part of the region's attractiveness), limited dependency on technology, and the dynamics of entrepreneurs. Given the beneficial effects on climate and environment, and the welfare produced, this is viewed as a 'good' example of a long-term development strategy. The challenges of global markets represent a major issue, but the region has shown noticeable resistance to date. In a broad sense, the tourism sector is both relatively stable and dynamic with regard to its inner adaptive capacity. However, stakeholders recommend that there should be formal linkages between high-level adaptation (e.g., at a regional level) and micro-level adaptation (e.g., at the level of individuals and companies).

2.4.6 Policy Recommendations

Stakeholders are central to the assessment of future needs for the formulation and recommendation of policies and strategies for adaptation. Stakeholders in the Tuscan workshop perceived a need for long-term policy strategies and for investment in technical training in order to adapt to climate hazards. In addition, the identification of 'tipping points', i.e., points where the magnitude of change due to climate change is such that the current management strategy will no longer be able to meet the objectives (Kwadijk et al. 2010) was considered a helpful approach for deciding when a management strategy might fail and for identifying alternative strategies required in a specific context. Participants postulated that the regional administration would find a local means of applying EU regulation to rural tourism, to ensure the preservation of both local traditional products (considered an economic asset and part of the cultural heritage of Tuscany) and food hygiene standards. Policy recommendations for Apulia include developing plans to produce more efficient infrastructure for water distribution and prevention of leaks, monitoring and regulated use of aquifers, and agricultural policies aimed at optimizing water use and irrigation. In the Valencia-Catalan case study, stakeholders support legislative changes that allow greater regional responsibility for improving coastal protection in the context of climate change. In a final example for the Tel Hadya case study (Syria), it is likely that the links established with policy makers over the course of the project will pave the way for changes in regulations governing the management and use of scarce water resources in the context of climate change.

2.5 Barriers to Stakeholder Participation and a Good Practice Checklist

Despite the substantial contribution of stakeholders to the CIRCE case-study assessments, there have been persistent obstacles to overcome. There were difficulties not only in indentifying the most appropriate set of stakeholders at a local and regional level, but also in encouraging more active participation. The barriers were largely due to a lack of resources (time and money); a lack of time commitment on the part of the stakeholders; and a shortfall in project finances. Stakeholder fatigue was a potential concern in some case study locations, as was the difficulty of motivating people in the absence of financial reward. Climate change is just one of a growing number of issues that have become a theme for public consultation. For example, the Apulia case-study team were unable to involve regional administrators and policymakers in consultation. The impediment was thought to be related to the overcrowded Italian political agenda at the time of the workshop. For some countries and politicians, direct involvement in environmental issues remains a low priority. Local organizations (e.g., consortia of private stakeholders) and politicians are reluctant to be involved until more targeted results are available. The challenge is to demonstrate in a clear and transparent manner why climate change should be of interest to a particular stakeholder and to exemplify the benefits of participation.

Based on awareness of these barriers, together with a review (i.e., what worked well? what was less successful?) of the stakeholder approaches and activities used (see Sect. 2.3) including the workshops (see Sect. 6.3.1), a good practice checklist for facilitating stakeholder involvement in research projects has been produced (Box 2.1). This is intended to be fairly generic and applicable to a wide range of projects related to climate, including those focusing on impact and adaptation assessments and the development of climate services. From the researchers' perspective, it is also concluded that another potential benefit of early stakeholder involvement is the opportunity to foster awareness and transfer information and knowledge to a wider audience through the stakeholders' own networks.

As Kasperson (2006) observed, the success of stakeholder participation exists not only in identifying the most appropriate players and exchanging information, but also in empowering them to act. The CIRCE experience suggests that a more participatory approach to adaptation planning for future climate change, which involves sharing of 'best' practice and the communication of accessible targeted information, will be fundamental to the successful development and implementation of adaptation strategies and capacity building in the Mediterranean region.

Box 2.1 Good Practice Checklist for Facilitating Stakeholder Involvement in Research Projects

- Ground-level knowledge and contacts are essential to ensure a network of the most appropriate stakeholders and key players.
- It is important to identify the relevant arena for decision-making – regional/local, and whether the focus is on practitioners or policymakers.
- It is good to include people from diverse perspectives: e.g., government departments and agencies, public and private sector bodies.
- Previous projects and meetings/conferences/workshops are useful for indentifying potential stakeholders.
- Identify at an early stage the level of stakeholder involvement that is desired (for the CIRCE case-studies, this was the 'participatory' level).
- For the participatory level of involvement, early contact with stakeholders is vital (ideally case-study workshops would have been held at the beginning of CIRCE as well as at the end).
- Email and telephone are useful for making first contact.
- Regular mailings are good for maintaining and developing contact, supported by informal/formal meetings.

(continued)

Box 2.1 (continued)

- It is important to provide motivation for stakeholders to get involved in the absence of financial reward or expenses for attending meetings etc. Thus the benefits and opportunities of getting involved need to be highlighted at an early stage (e.g., access to experts and data, sharing data/information, chance to influence the course of the research, working on 'real-life' problems and solutions together).
- In this context, it is good to identify very specific areas and stages of the research where stakeholders can contribute and participate.
- Demonstrate to stakeholders the 'added value' of outputs and products from the project, particularly where these are new.
- Aim to provide tailored/targeted and regionally/sectorally focused results and examples.
- Where necessary and appropriate, provide information in local languages.
- Briefing notes and information sheets focusing on 'What is it?', 'What does it show?' and 'Why is it relevant?' are good tools for presenting indicators of change and variability to stakeholders.
- Acknowledge that stakeholders may be as/more interested in current climate change and variability and shorter-term forecasts and predictions, rather than longer-term climate change projections.
- Clearly specify the aims and expected outcomes of stakeholder workshops.
- Circulate relevant material in advance of workshops, e.g., briefing notes and information sheets, leaflets, a perception or other preliminary questionnaire.
- Perception questionnaires are useful for beginning to understand stakeholders' attitudes towards risk and their decision-making framework.
- Encourage mutual feedback and exchange of information on areas or issues for which researcher input has had an impact on stakeholder decisions and activities, and areas for which stakeholder input has had an impact on research.
- Linking with other projects or events can help to increase the stakeholder workshop audience.
- Provide a feedback form at the end of workshops.
- The World Café is a good approach for facilitating active participation and discussion in a workshop setting.
- Involve stakeholders as speakers and, at a minimum, include a researcher/stakeholder 'roundtable' discussion in workshop programs.
- Use workshops to demonstrate and get feedback on tools developed in the project (e.g., Geographic Information Systems, decision-making tools).
- Consider using multi-media and audio-visual techniques where possible, particularly when involving the public.
- Provide stakeholders with feedback on final workshop outcomes, stakeholder relevant project reports, and policy recommendations and guidelines. This is especially important for encouraging follow-on work and active participation in future projects.

References

Benecke G (2010) The governance of clean development working paper series 007: stakeholder networks in carbon governance: the role of state-market relations in the Indian renewable energy sector. University of East Anglia, Norwich. http://www.uea.ac.uk/dev/gcd/Benecke+2010

Carter TR, Ones RN, Lu JX, Bhadwal S, Conde C, Mearns LO, O'Neill BC, Roun-sevell MDA, Zurek MB (2007) New assessment methods and the characterisation of future conditions. In: Parry ML, Canziani OF, Palutikof JP, van der Linden PJ, Hanson CE (eds) Climate change 2007: impacts, adaptation and vulnerability. Contribution of working group II to the fourth assessment report of the intergovernmental panel on climate change. Cambridge University Press, Cambridge, pp 133–171

Conde C, Lonsdale K (2005) Engaging stakeholders in the adaptation process. In: Lim B, Spanger-Siegfried E, Burton I, Malone E, Huq S (eds) Adaptation policy frameworks for climate change: developing strategies, policies and measures. Cambridge University Press, Cambridge/New York, pp 47–66

Hanson CE, Palutikof JP, Dlugolecki A, Giannakopoulos C (2006) Bridging the gap between science and the stakeholder: the case of climate change research. Clim Res 31(1):121–133. doi:10.3354/cr031121

Kasperson RE (2005) Bridging vulnerability science and practice. Paper presented at the 6th open meeting of the international human dimensions program, Bonn, Germany, 9–13 October 2005

Kasperson RE (2006) Rerouting the stakeholder express. Glob Environ Chang 16(4):320–322. doi:10.1016/j.gloenvcha.2006.08.002

Kwadijk JCJ, Haasnoot M, Mulder JPM, Hoogvliet MMC, Jeuken ABM, van der Krogt RAA, van Oostrom NGC, Schelfhout HA, van Velzen EH, van Waveren H, de Wit MJM (2010) Using adaptation tipping points to prepare for climate change and sea level rise: a case study in the Netherlands. Wiley Interdiscip Rev Clim Change 1(5):729–740. doi:10.1002/wcc.64

Moser SC (2008) A new charge: engaging at the science-practice interface. IHDP update (Newsletter of the Int. Human Dimensions Programme on Global Environmental Change), January 2008, pp 18–21

Moser SC (2009) Whether our levers are long enough and the fulcrum strong? Exploring the soft underbelly of adaptation decisions and actions. In: Adger NW, Lorenzoni I, O'Brien KL (eds) Adapting to climate change. Thresholds, values, governance. Cambridge University Press, Cambridge, pp 313–334, doi:10.1017/CBO9780511596667.021

Porter JR, Gawith M (1999) Temperatures and the growth and development of wheat: a review. Eur J Agron 10(1):23–36. doi:10.1016/S1161-0301(98)00047-1

Pretty JN, Guijt I, Thompson J, Scoones I (1995) Participatory learning and action: a trainer's guide, IIED training materials series no. 1. IIED, London, 268 pp

Vogel C, Moser SC, Kasperson RE et al (2007) Linking vulnerability, adaptation, and resilience science to practice: pathways, players, and partnerships. Glob Environ Change Hum Policy Dimens 17(3–4):349–364. doi:10.1016/j.gloenvcha.2007.05.002

Chapter 3
Physical and Socio-economic Indicators

Maureen D. Agnew, Clare M. Goodess, Debbie Hemming,
Christos Giannakopoulos, Skander Ben Salem, Marco Bindi, Mohamed
Nejmeddine Bradai, Camilla Dibari, Hesham El-Askary, Mutasem El-Fadel,
Mohamed El-Raey, Roberto Ferrise, José M. Grünzweig, Ali Harzallah,
Abdallah Hattour, Maria Hatzaki, Dina Kanas, Effie Kostopoulou, Piero
Lionello, Theib Oweis, Cosimo Pino, Basil Psiloglou, Salah Sahabi Abed,
Agustín Sánchez-Arcilla, Mohamed Senouci, Mohamed Zoheir Taleb,
and Annalisa Tanzarella

Abstract A set of physical and social indicators relevant to each Mediterranean case study has been developed within the context of the CIRCE case studies integrating framework. This framework approach provides a systematic means of structuring indicator selection and helps to provide a scientific basis for the assessment of climate-related impacts and vulnerability. A detailed set of criteria was developed

---

M.D. Agnew (✉) • C.M. Goodess
Climatic Research Unit, School of Environmental Sciences,
University of East Anglia, UEA, Norwich, UK
e-mail: m.agnew@uea.ac.uk

D. Hemming
Climate Impacts Analysis Team, Met Office Hadley Centre, Exeter, Devon, UK

C. Giannakopoulos • M. Hatzaki • E. Kostopoulou • B. Psiloglou
Institute for Environmental Research and Sustainable Development,
National Observatory of Athens, Athens, Greece

M. Bindi • R. Ferrise
Department of Plant, Soil and Environmental Science (DIPSA),
University of Florence, Florence, Italy

Institute for Biometeorology, National Research Council of Italy,
(CNR-IBIMET), Florence, Italy

C. Dibari
Department of Plant, Soil and Environmental Science (DIPSA),
University of Florence, Florence, Italy

S. Ben Salem • M.N. Bradai • A. Harzallah • A. Hattour
Institut National des Sciences et Technologies de la Mer – INSTM, Salammbô, Tunisia

A. Navarra and L. Tubiana (eds.), *Regional Assessment of Climate Change in the Mediterranean: Volume 3: Case Studies*, Advances in Global Change Research 52, DOI 10.1007/978-94-007-5769-1_3, © Springer Science+Business Media Dordrecht 2013

to select and refine indicators through an iterative process of review and consultation. Indicators represent key issues related to climate variability and change for each of the case-study locations. Seven key indicator themes are identified: climate and atmosphere; marine and coastal systems; terrestrial ecosystems and biodiversity; freshwater systems; agriculture and forestry; human health and well being; and, the economy. A number of core indicators are common to all case studies (for identifying common/disparate trends), others are common across generic case studies (urban, rural, coastal), and some are case-study specific. Data and methodological challenges in the indicator assessment included: data availability and quality limitations; distinguishing impacts from vulnerabilities, and climate from non-climate influences; and, identifying thresholds and coping ranges. Despite these difficulties, the selected set of indicators proved a useful and accessible tool for monitoring trends and portraying key information to regional stakeholders.

Keywords Climate change • Mediterranean • Indicators • Impacts • Integrated assessment

---

H. El-Askary
School of Earth and Environmental Sciences, Schmid College of Science and Technology,
Chapman University, Orange, CA, USA

Department of Environmental Sciences, Faculty of Science,
Alexandria University, Alexandria, Egypt

M. El-Fadel
Faculty of Engineering and Architecture, American University of Beirut,
Beirut, Lebanon

M. El-Raey
University of Alexandria, Alexandria, Egypt

J.M. Grünzweig • D. Kanas
Institute of Plant Sciences and Genetics in Agriculture,
The Robert H. Smith Faculty of Agriculture, Food and Environment,
The Hebrew University of Jerusalem, Jerusalem, Israel

P. Lionello
Centro Euro Mediterraneo per i Cambiamenti Climatici, CMCC, Lecce, Italy

Department of Material Science, University of Salento, Lecce, Italy

T. Oweis
International Center for Agricultural Research in the Dry Areas, ICARDA, Aleppo, Syria

C. Pino • A. Tanzarella
Department of Material Science, University of Salento, Lecce, Italy

A. Sánchez-Arcilla
Laboratori d'Enginyeria Marìtima (LIM/UPC), Universitat Politécnica de Catalunya,
Barcelona, Spain

S. Sahabi Abed • M. Senouci • M.Z. Taleb
Association Recherche Climat et Environnement – ARCE, Oran, Algeria

3.1 Introduction

Indicators provide a signal representing a complex message from potentially multiple sources in a rather simplified manner (e.g., OECD 1993, 2008; Hammond et al. 1995; EEA-JRC-WHO 2008; Esty et al. 2005). As such, they are a potentially useful and accessible monitoring tool for assessing trends and for portraying key information to policy and decision makers (Hák et al. 2007). In particular, projected changes in extreme climate indices are indicative of future changes in climate extremes (Peterson and Manton 2008). The CIRCE Mediterranean case-study indicators allow trends in a phenomenon to be identified at a regional or local scale, in this case as a response to changes in emissions of greenhouse gases, climate variability and change.

A set of indicators relevant to each case study has been developed within the context of the CIRCE Case studies Integrating Framework – CCIF (Sect. 1.3). The conceptual framework approach is recommended as a means of systematically structuring indicator selection and helping to provide a scientific basis (Niemeijer and de Groot 2008). Some of the indicators represent links between several nodes (the points of intersection that represent key variables used to make a decision) and between physical and social systems. For example, water resource indicators combine cross-sector information and are a particularly important indicator in semi-arid regions of the Mediterranean for resource management. Indicators are not combined to produce composite scores but are used to isolate and understand more clearly specific areas of impact and vulnerability. Some of the indicators are case-study specific (Sect. 3.3.2), some indicators are common across generic case studies (urban, rural, and coastal), and some core indicators are common to all case studies (Sect. 3.3.1) and provide a means of identifying common/disparate trends for sub-groups of case studies.

Building on the literature and local knowledge, key issues related to climate variability and change were identified for each of the 11 case-study areas using the structure of the CCIF. In addition, the key processes or activities (environmental and social) shaping vulnerability within the region or city, and their interrelationships were distinguished. An initial assessment of available datasets relevant to the case-study regions was undertaken in consultation with the wider project partners and regional stakeholders. Subsequently, the raw data for indicators were assessed for homogeneity, quality and credentials.

Physical and socio-economic case-study indicators have been identified for seven key themes: Climate and atmosphere; Marine and coastal systems; Terrestrial ecosystems and biodiversity; Freshwater systems; Agriculture and forestry; Human health and well being, and the Economy (Table 3.1). In addition, these individual themes are grouped into three broad categories: climate system indicators, biogeophysical system indicators, and social system indicators, as illustrated in the CCIF. The identification of indicator themes facilitates sector-based end-user research. This dual purpose approach to the presentation of Mediterranean indicators (i.e., by sector and by case study) was also applied in the dissemination of results (Sect. 3.2.5) through a series of structured thematic information sheets and web pages (http://www.cru.uea.ac.uk/projects/circe/).

Table 3.1 Organization of the CIRCE case-study indicators

System drivers	Indicator themes
Climate system indicators	1. Climate and atmosphere
Biogeophysical system indicators	2. Marine and coastal systems
	3. Terrestrial ecosystems and biodiversity
	4. Freshwater systems
	5. Agriculture and forestry
Social system indicators	6. Human health and well being
	7. The Economy

3.2 Methodology

3.2.1 Selection Criteria

The broad terms of reference for the integrating case-study indicators were that they should be appropriate to the Mediterranean (in terms of climate variability and change, and vulnerability); and able to be constructed relatively easily using reliable and appropriate data sets for the present-day and the future, and where appropriate drawing on outputs from other CIRCE research lines. A detailed set of criteria, adopted from Erhard et al. (2002), and grounded within the context of the CCIF was developed for case-study indicator selection:

- Is the indicator sensitive to climate change (are there changes/trends in the indicators that have a demonstrable association with climate variability and change)?
- Is the indicator relevant to and representative of the Mediterranean case-study region?
- Is the indicator relevant to the scope of the proposed climate change impacts adaptation and vulnerability case-study assessment?
- Is the length of the data time series (upon which the indicator is based) sufficient to capture the response to climate change?
- Does the data have adequate spatial resolution? The desired resolution is regional or sub-regional for the rural and coastal case-studies, and city-wide or district for the urban case studies.
- Is the indicator relevant to policy makers and stakeholders?
- Does the indicator have resonance – is it accessible and readily understood and interpreted by decision-makers and stakeholders?
- Quality of the data: is it well-founded with acceptable credentials?
- Data homogeneity: is there a change in definition or methodology for measuring or collecting the data over time?
- Is the raw data updated at regular intervals of time?
- Is the indicator suitable for incorporation into a numerical integrated impact assessment model?

In practice, the availability of data relevant to the regional case studies was a major defining criterion, and in some circumstances, due to necessity, the selection criteria were relaxed and indicators were presented with caveats (see Sect. 3.2.4). For example, although the length of the data series was, in the case of marine data, insufficient to retrieve long-term trends, this information was considered alongside wider datasets. Data of sufficient length and continuity were not always available for the specific vulnerability issues identified (e.g., water use by sector) but what data were available were considered together with local knowledge and information. For some case studies, data homogeneity could not always be confirmed due to extenuating local circumstances (e.g., in the case of observed climate in Beirut where monitoring has been hampered by civil unrest). In this particular case, indicators were developed using available station data and compared to observed gridded data available for a wider area.

3.2.2 Reviewing and Refining Indicators

The process of systematically selecting the final set of indicators involved several iterations of review and consultation. Indicators were chosen from the revised indicator pool according to the degree of relevance to the particular case-study location and the availability of appropriate data. In addition, a smaller set of core indicators were selected that are common to (i) each of the generic case studies and to (ii) all case studies. Through this process, the overall objective was to build a structured set of case-study specific indicators applying formal selection criteria, grounded in a conceptual framework that allows vulnerability to climate variability and change to be assessed at a regional/local level, across generic case-study sites and within the context of the wider Mediterranean area. In practice, the process was highly constrained by data availability and project resources (in particular, time).

Indicators were calculated from observed or modeled case-study data, initially for past and present periods to identify any observed trends in climate and assess present-day climate vulnerability and impacts. Subsequently where model data were available (i.e., climate and marine, and a smaller number of biogeophysical and social indicators), indicators were constructed for future time periods.

3.2.3 Assessment for Trend, Thresholds and Coping Range

Indicators were assessed for changes and trends over time where data (quality and series length) permitted. Thus for the observed climate and marine data, anomalies from the long-term average (1971–2000) were computed and linear (regression) trends were assessed for statistical significance (using the Mann-Kendall test). Where long-term trends where present they were expressed as a unit change in measurement (positive or negative) per decade. Rates of climate change are particularly important in terms of the environmental and social response. Confidence in the trend was assessed using the derived likelihood value categorized according to IPCC

2007 terminology, 'virtually certain', 'extremely likely', etc. (Risbey and Kandlikar 2007). Biogeophysical and social indicators were handled somewhat differently to climate indicators. Change and trends over time, for example, were only assessed where there was a sufficient quantity (and consistent quality) of long-term data. In cases of data monitoring infrequency, values of the indicator at two or more points in time were compared, as for environmental monitoring of waste water discharge in the Bay of Oran (Table 4.5) and land cover surveys in Tuscany.

Long-term changes in the indicators are not always linear; some are strongly non linear (see Sect. 3.5.2 for further information). In addition, a number of the indicators show a change in the rate of increase in recent decades. For example, many of the temperature indicators show a faster rate of increase from the 1970s onwards (e.g., maximum temperature in Athens, Beirut, the Gulf of Oran, the Gulf of Gabès, and the Judean Foothills). This acceleration in the rate of warming for 'hot' extremes is observed in independent data sets and can be considered quite robust (Part I, Sect. 2.3.2).

A threshold is the point at which further stress (e.g., continued warming) will produce a non-linear response in an exposed biogeophysical or social system or activity. *Systemic* thresholds (e.g., the plant wilting point during drought) can be defined objectively and have particular utility in assessing change. *Impact* thresholds identify the point at which a climate impact is regarded as 'unacceptable' (Kenny et al. 2000; Jones 2001), are used to assess risk and are subjective and commonly value-laden (Carter et al. 2007). Thresholds in climate impact response have been identified from case-study data, literature, or through stakeholder consultation. However, they are viewed in a differing sense among the case-studies. For example, they tend to be viewed as the point of system collapse by the coastal case studies (breach of coastal defenses); whereas in other case studies they are viewed as the point at which a significant impact is experienced. For example, in the case of energy consumption in Athens, the threshold is the thermal conditions for which space heating or cooling is required (Fig. 4.3). While thresholds are readily identifiable for some indicators, they are not so apparent in others. For example in the Judean Foothills, a positive linear association is shown between the stem volume of Aleppo pine and mean annual rainfall sampled along a moisture gradient (Sect. 4.3.3; Fig. 4.8). Although there are insufficient observed data for threshold detection, there are present-day drier analogue locations where pine is not recorded, providing evidence of a lower rainfall threshold below which pine growth is not sustained. In the Judean Foothills, the threshold for growth of Aleppo pine planted on hill slopes is around 200 mm of mean annual precipitation for a substrate with rocky outcrops and is approximately 250 mm for loess conditions without outcrops. For successful recruitment of new trees in sustainable pine forests, higher amounts of rain of around 350 mm are required.

Lower and upper thresholds for an impacted system or activity (represented by the chosen indicators) define the coping range for which variations in climatic conditions can be accommodated (Smith et al. 2001; Carter et al. 2007). The coping range is viewed as a dynamic phenomenon that can be both constrained and extended through a complex system of pressure and adaptive mechanisms in physical and social dimensions (Füssel 2007). The CIRCE case-studies have identified socio-environmental factors which may expand the coping range (such as an increase in coastal protection and management, supplemental irrigation, heat wave early warning systems) or narrow the coping range (such as environmental degradation, and increasing pressures of population growth). It is suggested that behavioral adaptation measures such as improved hygiene and sanitation could limit or reverse projected future adverse disease outcomes in some case studies, such as Beirut, and increase human health resilience (see Sect. 6.3.2.2).

An analysis of past extreme climate events serves as a useful indication of the potential impacts of future climate change and an opportunity to learn from previous adaptive strategies and coping mechanisms. The Athens urban case study, for example, assessed the impacts of the exceptionally hot and dry summer of 2007 on fire occurrence and social well being (Box 5.1). In Tuscany, the year 2003 was notably the driest and warmest year recorded in the past 30 years. Several negative impacts were recorded for the region, the lowest mean annual wheat yield, the greatest negative water balance (ranging from −200 to −500 mm; Fig. 3.2), and the highest peaks in forest fire events (1,036 fire events covering 4,130 ha; Fig. 3.3) and water consumption for agriculture (Fig. 3.9). In the Judean Foothills, the 1998–2000 drought caused widespread mortality in planted forests, with up to 20% of the forest area affected (JNF 2002). Tree mortality varied according to tree species and stand characteristics, such as density, age and geophysical properties.

3.2.4 Data and Methodological Challenges

The selected set of indicators (discussed further in Chap. 4, and included in Appendix 2) represents a complex interacting set of processes operating on a variety of temporal and spatial scales. Using contextual information from literature, stakeholder dialogue and the structure of the conceptual framework, the research teams sought to identify the most pertinent issues for each case study in relation to current and future climate states. However, limitations and weaknesses in the data and methodology exist, and the real-world complexities cannot be fully captured by a restricted set of indicators.

It has been difficult to achieve a truly balanced set of indicators for each case study and across the three main systems (climate, biogeophysical, social) due to limitations in the availability of data of sufficient quality and length at the appropriate scale of measurement. As a result, the state and nature of knowledge varies across the physical and social domains, and between individual case studies. It is therefore possible that there is a key characteristic of vulnerability that has not been recognized due to absence of information.

For some case-study locations such as Beirut, particular data quality issues have arisen due to periods of civil unrest during which data are either unavailable or of insufficient or uncertain quality. For some regions, there is a shortage of structured environmental monitoring, such as in Alexandria and the West Nile region, the Gulf

Table 3.2 Changes in the diversity of vegetation types ('+' present; '−' absent) in natural ecosystems with changes in mean annual precipitation

Vegetation type	Mean annual precipitation (mm)			
	300	402	411	481
Annual herbaceous land	+	−	+	−
Perennial herbaceous land	+	+	+	−
Dwarf shrubland, dwarf shrubland-savanna	+	+	+	+
Shrubland	−	+	+	+
Open forest, parkland	−	+	+	+
Maquis	−	−	+	+

Data source: summarized from Nir Herr (personal communication, 2010)
Categories of vegetation types are a generalization of the more detailed vegetation classes developed by the JNF-KKL

of Oran, Beirut, and Tel Hadya. In other regions, environmental monitoring programs have been established but the length of time for which data are available is too short to allow identification of long-term trends and characterization of the response to climate variability and change. This is particularly true with respect to marine data (such as significant wave height, relative sea level, and storm surges) for the coastal case-study regions. The relative sea-level series for Sfax Harbour, Tunisia, for example, covers only 9 years (the most complete in the area), however, the estimated decadal rate of increase of 2.6 cm/decade is close to the rate of 2.1 cm/decade calculated for the Mediterranean basin for the longer period 1992–2005 and based on altimetry data (Criado-Aldeanueva et al. 2008).

A particular methodological challenge is in combining and using indicators derived from both quantitative and qualitative data. This issue is most apparent in, but not limited to, interactions between environmental and social dimensions. More qualitative or survey data is common with respect to vulnerability assessments of terrestrial and marine ecosystems within the Mediterranean case studies. In the Judean Foothills, for example, survey presence/absence data were used to describe the range of vegetation types along a rainfall gradient (Table 3.2), and were combined with more quantitative information on the stem volume of Aleppo pine (Fig. 4.8).

The distinction between impact indicators and vulnerability indicators is not always clear in the case studies. For example, saline intrusion is a particular issue of concern in the West Nile Delta, and may be considered as a response to climate-induced sea-level rise and an impact indicator in its own right. However, freshwater availability is also a concern in the region and is vulnerable to higher temperatures, lower rainfall, drought, and salinization. Therefore in this context, saline intrusion represents an indicator of vulnerability.

There are also issues of scale (temporal and spatial) which add to the complexity and challenges of the integrated case-study assessment. For example, indicator data are available at a variety of spatial scales (station observations, gridded data, administrative regions, and areas defined by land use such as forest plantations) and temporal scales (e.g., annual, monthly, daily, hourly). For each case-study, the most appropriate means of linking different scales was sought. In the case of the Gulf of Gabès and West Nile Delta case studies, a Geographic Information System (GIS) was found to be a useful means of representing and overlying spatially referenced layers of data (e.g., Fig. 2.2).

Within the same case study, some of the impact indicators provide statistical evidence of association between climate system drivers and the physical and social dimensions; while other linkages are explored and discussed in a more qualitative sense. Although the integrated assessment of vulnerability is rather limited in predictive capacity because some dimensions are less clearly defined (in a statistical sense), the interpretative and instructive value is the greater strength. For example in the Tuscany rural case study, a statistically significant negative association was found between wheat yield and mean annual temperature ($r = -0.77$; $p < 0.01$), and a positive association with total annual precipitation although weaker and not statistically significant ($r = +0.47$; $p < 0.1$). In contrast, although there is limited information on water availability in Tuscany (Fig. 3.9), annual water consumption by the agricultural sector is available for the years 2000–2008, and the highest peak in consumption occurred in the year 2003 (one of the driest years on record) when agriculture consumption reached 180 million m^3. During the same year, a water deficit was recorded commensurate with a fall in mean annual wheat yield (Fig. 3.2). Although available data on water consumption were not sufficient to build statistical models, the information provided deeper insight into regional aspects of climate vulnerability.

Issues relating to indicator data and methodological challenges are discussed further in the context of assessment uncertainties in Sect. 5.4. Many of the research gaps and needs identified in Sect. 6.6 also relate to these issues.

3.2.5 *Methods of Presentation*

The first two series of thematic information sheets (accessible from each of the case-study home pages, http://www.cru.uea.ac.uk/projects/circe/index.html) were created around key sets of indicators for each case study. The first series of information sheets was focused on observed climate and marine indicators, and the second series focused on indicators representing key aspects of current biogeophysical and social vulnerability in the case studies. These information sheets were targeted at the user-community and attempted to communicate key indicator messages using a clear and accessible format. Each indicator was presented using a common structure, a figure (showing spatial or temporal trends in the indicator) and textual information given under a three-fold set of sub-headings: 'What is it?', 'What does it show?' and 'Why is it important?' Overview observed climate information sheets were also provided covering the generic urban, rural, and coastal case-studies. These sheets highlight cross-study similarities and dissimilarities for each of the groups of generic case studies.

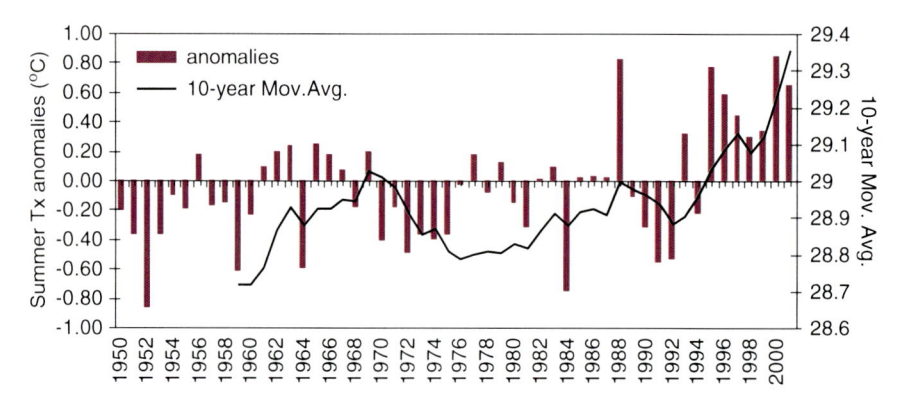

Fig. 3.1 Beirut Summer maximum temperature (Tx) anomalies (°C) from the 1971 to 2000 average (*left axis*) for the Aéroport International de Beyrouth (AIB) station (*bars*). 10-year moving average of the absolute summer Tx (*right axis*) for the AIB station (*line*)

Visually, indicators were presented as (i) time series plots, and typically for the climate indicators, as anomalies from the long-term trend (e.g., summer maximum temperature in Beirut, Fig. 3.1); (ii) spatial plots which show regional variation in trends in, for example, evapotranspiration in Tuscany (Fig. 3.2); (iii) combination spatial – temporal plots, as for the time series representing the number of Tuscan fires/area burnt presented with an inset plot showing the distribution of area burnt by Tuscan administrative municipality (Fig. 3.3); and (iv) pie-charts or bar charts used to portray changes in the percentage composition of, for example, land use, vegetation type, or water consumption.

3.2.6 Integrated Vulnerability Assessment

For each case-study, potential indicators were mapped using a schematic flow diagram based on the CCIF framework. Figure 3.4 shows an example for the Judean Foothills. The indicators are broadly organized in the diagram to represent climate (blue), biogeophysical (green) and socio-economic (orange) systems. The Judean Foothills case-study location straddles a steep rainfall gradient between an annual total of 300 mm in the semi-arid south to 550 mm in the dry sub-humid north. The rural economy is largely dependent on agriculture, for which water resources and availability are the main limiting factors. Potential impact pathways are direct through physiological processes and indirect via the effects on water resources. The key climate hazards are the frequency and intensity of extreme events such as drought and heat stress. Biological activity in natural ecosystems alters in response to moisture changes at all levels, from leaf-scale photosynthesis to plant growth (Fig. 4.10), to landscape-scale vegetation composition (Table 3.2) and carbon sequestration. These climate-driven changes in ecosystems are expected to significantly affect forestry (e.g., pine stem volume) and agriculture (e.g., rain-fed wheat yields),

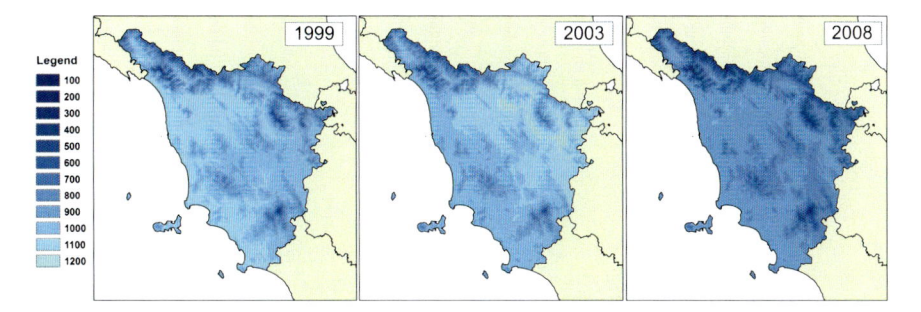

Fig. 3.2 Annual evapotranspiration (mm/year) in Tuscany for the years 1999 (average precipitation), 2003 (dry year), 2008 (wet year) (Source: Interdepartmental Centre of Bioclimatology, University of Florence, data processed by Dr. Marco Napoli)

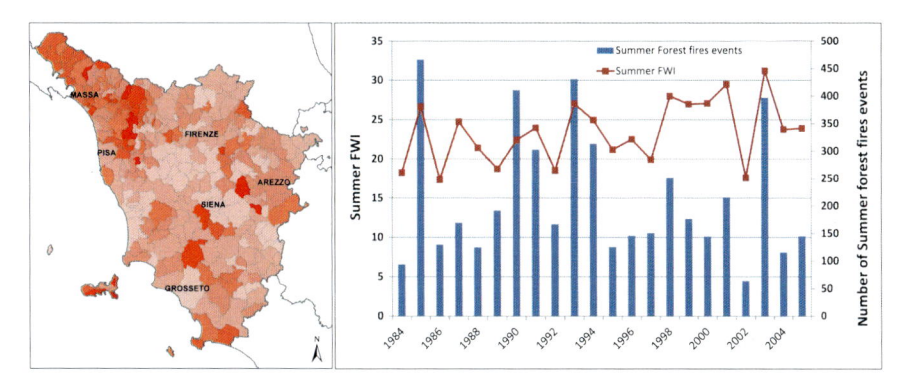

Fig. 3.3 Administrative municipalities most affected (*deepest red*) in terms of total area burnt (*left*). Fire Weather Index (*red line*) and number of summer fire events (*blue bars*) from 1984 to 2005 (*right*) (Source: Moriondo et al. 2006, Administrative Region of Tuscany, Wild Fire Service)

tourism (e.g., visits to national parks) and the wider rural economy, and may ultimately influence policy response.

For the Oran case study, work focused on building a common framework which included a set of quantitative indicators, such as population dynamics, urbanization, urban land use, surface temperature and carbon uptake, for developing vulnerability indices for the Mediterranean basin. The interactive stakeholder approach adopted in this case study constituted an important step towards the development of a more comprehensive integrating framework applicable to all Mediterranean coastal regions for assessing future climate scenarios and their cross-sector impacts in these fragile coastal zones.

The Mediterranean case-studies approach views the integrated assessment as a conceptual framework comprising key indicators representing driver, nodes and interactions within the environmental-social-economic system. It is not a computer simulation model. The case-study indicator linkages diagram (e.g., Fig. 3.4) proved a useful preparatory stage for structuring the assessment, and is viewed and used as an important research tool for raising scientific questions. It also proved a valuable

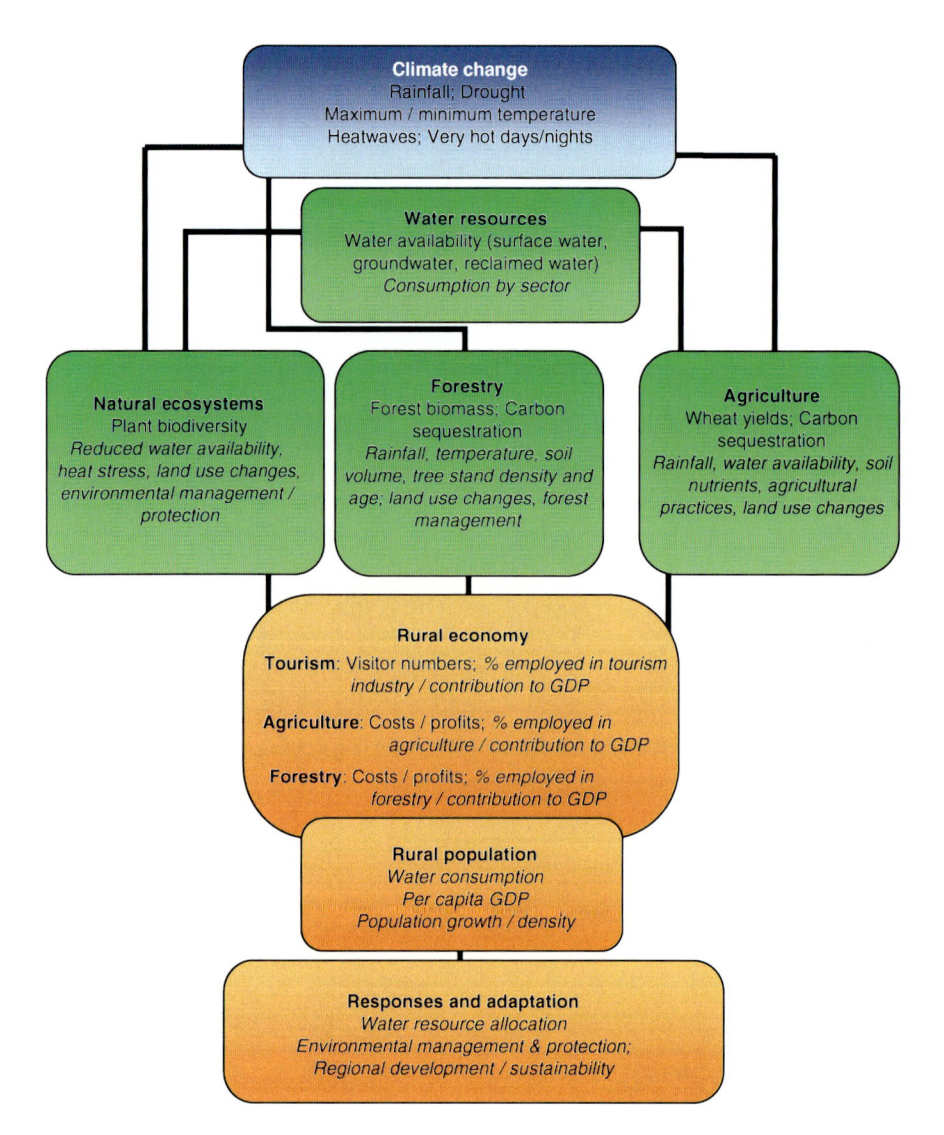

Fig. 3.4 Indicator linkages for the Judean Foothills. The *blue box* denotes key climate drivers, the *green boxes* denote key biogeophysical impacts and vulnerabilities (*italics*); the *orange boxes* denote key social impacts and vulnerabilities (*italics*)

visual communication tool for discussing issues of vulnerability and adaptation with regional stakeholders, allowing identification of critical types and sources of information uncertainty (Sect. 5.4) and knowledge gaps (Sect. 6.6). Within this approach, the description of cross-sector linkages and interactions and feedbacks between climate and non-climate drivers has the flexibility to include both quantitative and more qualitative information.

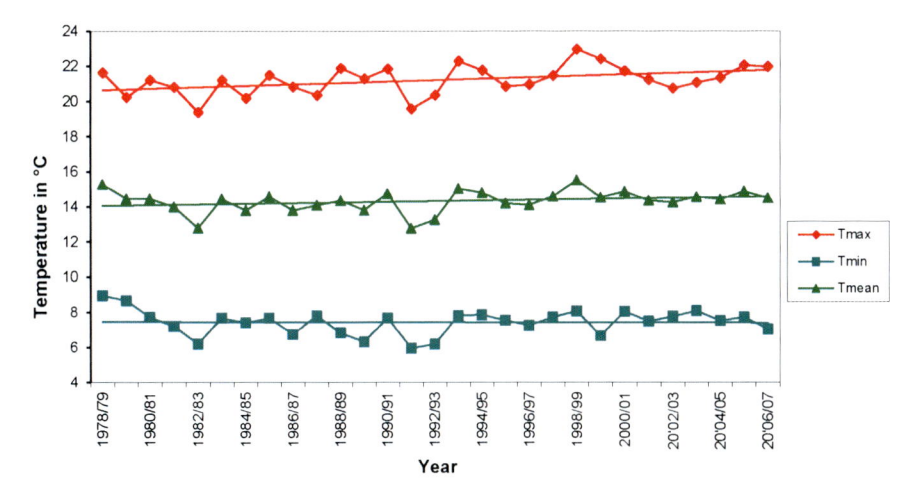

Fig. 3.5 Mean growing season (October to June) daily maximum, minimum and mean temperature (°C) for Tel Hadya, Aleppo-Syria, for the period 1978/1979–2006/2007

3.3 Climate and Atmosphere Indicators

Variability and change in the climate and atmosphere are viewed as fundamental drivers of the dynamics in biogeophysical and social systems (CCIF, Chap. 1, Fig. 1.2). Therefore, the climate and atmosphere indicators (Appendix 2) form the basis of the integrated assessments of climate change impacts, adaptation and vulnerability in the Mediterranean case studies, and the subject of the first set of information sheets (Sect. 3.2.5). The climate indicators are sub-grouped into (i) core climate indicators, and (ii) case-study specific climate indicators.

3.3.1 Core Climate Indicators

The core climate indicators are relevant to all case-study locations and comprise:

1. Maximum temperature for a key season appropriate to the case study, and
2. Precipitation for a key season appropriate to the case study.

Climate indicators are expressed as anomalies, the deviation from a common baseline period (1971–2000 in the case of observations) where the instrumental record is sufficiently long – as in the example of maximum summer temperature for Beirut (Fig. 3.1; see also Figs. 4.4 and 4.5). In other examples, the absolute values are used such as mean growing season temperatures for Tel Hadya (Fig. 3.5).

Climate indicators were also derived for a common future period (2021–2050) using daily climate model output for a number of fields including: total precipitation, temperature (mean, maximum and minimum), net surface solar radiation, relative humidity, maximum wind speed, total cloud cover, surface evaporation, sea surface temperature, sea level, and daily sea-surface salinity. Climate model data were derived from another CIRCE research line (details are given in Part I, Chap. 3) and are discussed in the context of the case studies in Sect. 5.2. Six climate scenario simulations for the Mediterranean region and the A1B emissions scenario were performed using a range of coupled atmosphere–ocean models at spatial resolutions of up to 30 km for the atmospheric models. All future changes are expressed relative to the scenario baseline period of 1961–1990 (used across the CIRCE project) in order to minimize effects of model bias (Sect. 5.4). Examples of indicators constructed for the case studies using future climate projections are presented in Chap. 5.

3.3.2 Additional Case-Study Specific Climate Indicators

Additional climate and atmosphere indicators were selected for each case study according to the specific issues of relevance. These case-study specific indicators included other aspects of average climate such as minimum and mean temperature, and climate extremes such as hot days (Fig. 4.4), hot nights, heat wave duration index, drought indices (e.g., standardized precipitation index, self-calibrated Palmer Drought Severity Index), greatest 3-day rainfall, the tourism season (expressed as the length in days based on thermal considerations), and the aerosol optical depth (AOD). AOD is a measure of the opaqueness of air and an indicator of the relative amount of aerosols suspended in the atmosphere (Fig. 3.6).

The urban case studies (Sect. 4.2.2) have generally been concerned with aspects of climate related to heat stress and heat waves (relevant to human health and energy consumption), and precipitation extremes, such as intense rainfall associated with flash floods in urban areas, and periods of drought increasing the risk of peri-urban forest fires. The rural case studies (Sect. 4.3.2) have selected climate indicators relevant to crop yields and natural ecosystems (drought indices, heat stress, and seasonal rainfall), while the coastal case studies (Sect. 4.4.2) have chosen climate indicators reflecting important coastal zone sensitivities (e.g., a temperature-defined tourist season, hot days, and wave storm frequency).

3.4 Biogeophysical Indicators

3.4.1 Key Themes

Biogeophysical indicators are grouped under four themes (Table 3.1): Marine and coastal systems (e.g., sea surface temperature, relative sea level, wave height, marine pollution, alien marine species, fish stocks, coastal erosion, and salt-water intrusion); Terrestrial ecosystems and biodiversity (e.g., fire frequency and area

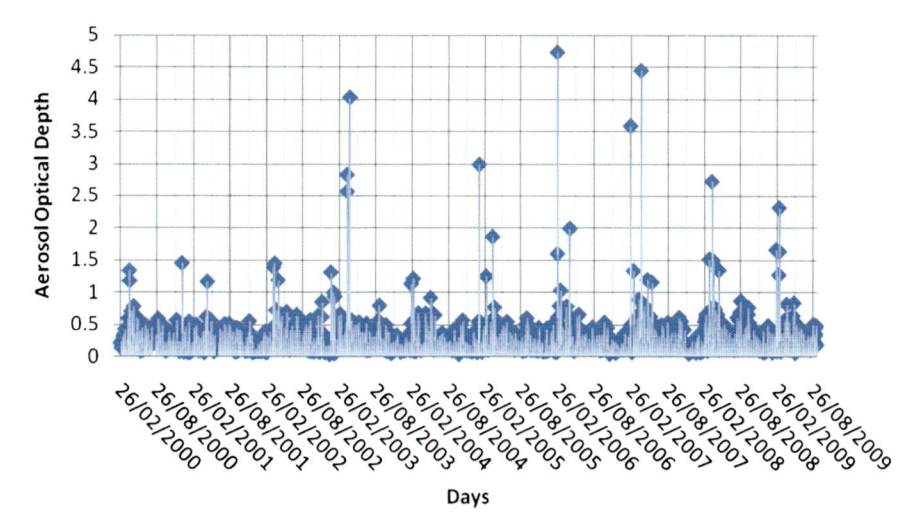

Fig. 3.6 MODIS satellite sensor derived level 2 (10×10 km) daily aerosol optical depth (AOD) over Alexandria, Egypt during the period 2000–2009 (El-Askary et al. 2009)

burnt, ozone exceedance, land use changes, forest biomass and tree health, vegetation types, and carbon sequestration); Freshwater systems (e.g., water availability, and evapotranspiration); and Agriculture and forestry (e.g., yields of key crops, plant ecological surveys, and woody stem volume). The selected indicators are listed by key theme and case study in Appendix 2.

3.4.1.1 Urban Case-Study Themes

Biogeophysical indicators are considered in the urban case studies although their importance is less of an end-point impact indicator and more of an indicator of vulnerability for social systems. For example in Beirut, the negative effect of climate variability and change on the availability of clean drinking water is a critical vulnerability issue, with the impact end point human morbidity and mortality. The key biogeophysical indicators vary between urban case study: freshwater quality and availability in the case of Beirut; air quality and peri-urban fires (Figs. 5.12 and 5.13) in the case of Athens; air quality, availability of freshwater, and sea-level rise in the case of Alexandria (see Sects. 4.2.1 and 4.2.3; Fig. 4.1).

3.4.1.2 Rural Case-Study Themes

Crop yields (particularly wheat) are a focus of all the rural case studies (see Sects. 4.3.1 and 4.3.3; Fig. 4.6). An impact-led model of vulnerability has explored the sensitivity of yields to CO_2, temperature, and the availability of water. The optimal range in climate and environmental conditions is known to vary according to crop species and growth stage. Adaptation and coping mechanisms considered in the rural case studies include choice of cultivar, changing plant and harvesting dates, and supplemental irrigation. Farmer income can be supplemented through rural tourism, but there is a risk to tourist attractiveness (especially in Tuscany) if the landscape changes due to the withdrawal of traditional crops in response to climate change.

3.4.1.3 Coastal Case-Study Themes

Among the four coastal case studies (see Sects. 4.4.1 and 4.4.3; Fig. 4.11), key biogeophysical indicator themes are closely associated with the impacts of sea-level rise and changes in storm severity on the littoral zone (e.g., coastal erosion and sea-level intrusion) and marine ecosystems (e.g., sea-water temperature, invasion of alien species (Fig. 3.7), and marine biodiversity). Each coastal case study has specific issues of regional importance. For the Gulf of Valencia – Catalan coast, relative sea level and coastal erosion, marine water quality, and saline intrusion; for the Gulf of Oran this is the fragile but ecologically important coastal nature reserves and its exposure to pollution; and for the Gulf of Gabès this is the vulnerability of low-lying tourist resort islands to sea-level rise and beach erosion together with a decline of traditionally important fish stocks. Finally, for the West Nile Delta, key risks arise from the multiple threats of environmental degradation and sea-level rise.

3.4.2 Key Challenges

The availability of appropriate data (in terms of length, quality, and frequency of measurement) has limited the knowledge base in some case studies. The identification of marine hazards (such as storm surges and sea level rise) is particularly constrained by sampling size and length of time series (Sect. 3.2.4). Therefore where possible, case-study data were also compared to larger data sets available for the Mediterranean Basin, such as in the case of temperature and precipitation extremes (Chap. 5). There is a paucity of long time series of ecological data. In the Gulf of Oran, marine biological impact indicators (such as invasion of alien species) were represented by a combination of single year surveys of species percentage coverage, fishermen observations, prey species stomach contents, and marine discharge data covering two comparative years (Table 4.5), while vulnerability indicators were illustrated by 5 months of water chemistry for an estuarine and freshwater site.

There are considerable challenges in distinguishing between changes in biogeophysical indicators that are a response to climate and those that are due to other single or combined environmental or social system drivers. For example, significant shoreline changes have been observed around the Gulf of Gabès from satellite images. However, it is unclear to what extent these changes are in response to human activity (construction and industrial works in the coastal zone) and how much can be attributed to climate-moderated changes in the marine environment. Similarly, in

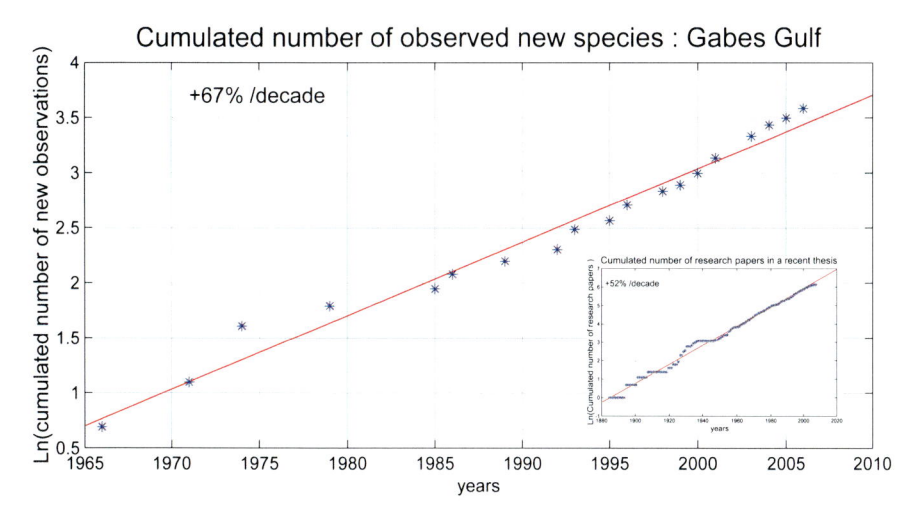

Fig. 3.7 Cumulated number of observed alien species (logarithm) in the Gulf of Gabès per observing year 1966–2006; the *solid line* is the linear trend. First records are based on a compilation of the observed alien species (40) from available papers, thesis and reports. The *inner plot* shows the increasing number of studies on alien species (to remove the bias introduced by the increased research, and hence reporting, activity)

Beirut a common response to alleviate water shortages associated with increased demand due to population growth and social development has been to rely on groundwater extraction thus inducing salt-water intrusion. The latter is also expected to increase as a result of sea-level rise induced by climate change. The assessment of the 'additional' factor that can be attributed to the impacts of climate change remains a key challenge. The difficulty in isolating a climate signal is also particularly apparent in the Tuscany case study. While wheat yields are sensitive to climate conditions, resultant yields are strongly influenced by farm management (e.g., input of fertilizer, irrigation, technical resources, choice of cultivar) and European, national and regional policies. In Tuscany, the stable wheat yields recorded in the recent period 2004–2008 are likely due to recent Common Agricultural Policies, through which Tuscan farmers have more individual discretion to meet market demands in terms of 'quality', and are encouraged to apply agricultural practices and investment capable of guaranteeing a more stable and higher crop output.

3.5 Social Indicators

3.5.1 Key Themes

Social indicators are subdivided into two sub themes: human health (including all-cause mortality, and water-borne disease); and economic systems including the tourism sector (daily climate tourism potential index; socio-economic tourism losses, tourism arrivals, and visits to a national park), the energy sector (production/consumption by energy type and sector), fisheries and aquaculture (fish catch/landings), agricultural production (e.g., olive oil and wine), and the water industry (e.g., water consumption by sector).

3.5.1.1 Urban Case-Study Themes

Among the three urban case studies, key social indicator themes are associated with impacts on human health and are closely connected to the driving biogeophysical indicators (see Sects. 4.2.1 and 4.2.3; Fig. 4.2). In Athens, air pollution and heatwaves act synergistically to increase mortality rates in summer (Fig. 3.8; and see Sect. 5.3.3.1 for further information). In Beirut, lack of access to clean water resources is a health issue affecting morbidity and ultimately mortality. In Alexandria, poor air quality (Fig. 3.6) from pollution and dust storms has a significant human health effect. Additional social indicators for each urban site were also considered. For example in Athens, energy consumption levels throughout the year are closely related to ambient temperatures (Fig. 4.3) as communities attempt to maintain a comfortable living and working environment.

3.5.1.2 Rural Case-Study Themes

Due to time and data constraints, a fully integrated and quantitative assessment of all aspects of climate vulnerability for each of the rural regions was not possible. Instead, research focused on key linkages between the main impact indicators: agricultural production (such as, wheat yield in the Judean Foothills; Fig. 4.8), tourist arrivals, and forest fires (Fig. 3.3) and vulnerability indicators: water availability (Fig. 3.2), land-use change, and population change of relevance to rural climate-sensitive activities (see Sects. 4.3.1 and 4.3.3; Fig. 4.6).

3.5.1.3 Coastal Case-Study Themes

Among the coastal socio-economic themes considered (see Sects. 4.4.1 and 4.4.3; Fig. 4.11), the tourism sector is identified as an important and rapidly expanding economic system for much of the Mediterranean coastline. A theoretical simulation was used as a tool to investigate the socio-economic impacts related to climate change (warming coupled with increased beach erosion) in the Gulf of Gabès case study (Fig. 5.23). Fishing and aquaculture is another key economic sector in coastal zones, affected by the vulnerability of marine ecosystems to the collective pressures of sea warming and marine pollution as highlighted in the Gulf of Oran and the Gulf of Gabès case studies (Fig. 5.26). Climate-induced salt-water intrusion was a particular theme for the Gulf of Valencia and the West Nile Delta with wide-reaching societal impacts on freshwater resources, agriculture, settlements and land use.

3.5.2 Key Challenges

- Social indicators, like biogeophysical systems, usually respond to a complex combination of environmental and social drivers, but in the case of social indicators the causal chain is typically longer, and there are greater interactions with non-climate drivers. For example, tourism arrivals/visits (considered in the Gulf of Gabès, Judean Foothills, Apulia and Tuscany) are a behavioral response to a complex array of factors including climate conditions in the sending and receiving region, the general attractiveness of the resort (partly attributable to climate but also to service levels and facilities), currency exchange rates, ease of access, cost of travel (which may reflect climate policies), civil and political stability, and current fashions and trends in holiday destinations. Circumstances, such as the ongoing political situation in Algeria, Tunisia, Egypt and Syria (at the time of writing), can become the dominant system driver, temporarily diminishing the climate signal (see Sect. 6.4.1 for further discussion).
- Social indicators may respond in a non-linear fashion to climate variability and change. For example, in the case of excess deaths and daily temperature in Athens, deaths tend to rise as temperatures increases above a threshold of 34°C (Fig. 3.8). Between 23 and 33°C no excess summer deaths were recorded. A further complication is that health vulnerability to critical temperature thresholds is not socially homogenous, but varies according to age, pre-existing illness (affecting individual resilience to heat stress) and social class (affecting access to health care, air conditioning, and general living conditions). In addition, thresholds may shift in the future due to physiological adaptation as the body acclimatizes to a changed climate (Sect. 5.3.3.1). In the social context, thresholds tend to be case specific, founded on subjective assessments of what constitutes acceptable risk and impact, and are influenced by cultural factors (Meze-Hausken 2008).
- The nature of socio-economic data has intrinsic challenges, tending to shorter time series; infrequent/irregular measurement; changes in methods or frequency of recording data over time; and data that are more qualitative in nature. Annual data on water consumption by sector in Tuscany (Fig. 3.9) for example, are available for a period of 14 years (1995–2008); while industrial water consumption covers 13 years (1995–2007), water consumption by the agricultural sector covers a shorter 10-period (2000–2009), and domestic water consumption is only available for 1 year, 2006. These regional data inadequacies highlight the importance of consistent monitoring of social and biogeophysical indicators to highlight key trends in vulnerabilities and impacts (Sect. 6.6).

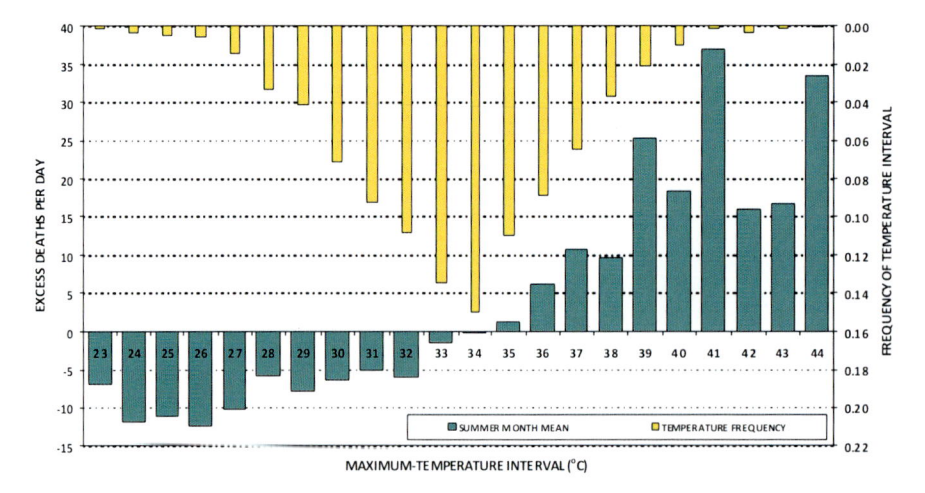

Fig. 3.8 Daily excess summer deaths in Athens by maximum air temperature interval (*blue bars*) for the years 1992–2006. The frequency of occurrence of each temperature interval is shown using *yellow bars*

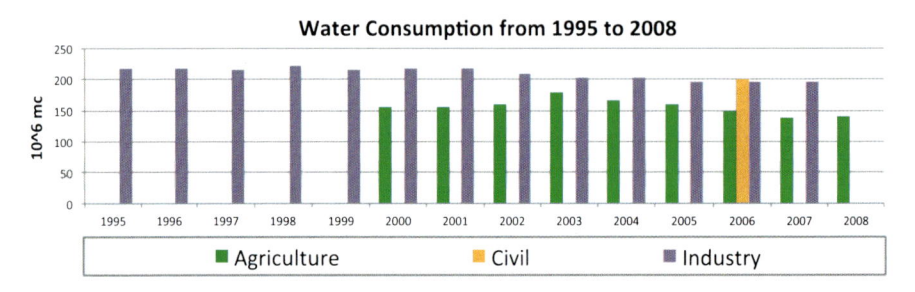

Fig. 3.9 Water consumption (million m³) in Tuscany by agriculture (2000–2008), civil (2006 only) and industry (1995–2007) sectors (Source: Regional Tuscany public administration)

- Spatial scale differences between social and environmental data present methodological challenges. Observed climate and other environmental data frequently relate to a specific observation/sampling site, whereas socio-economic data may relate to an administrative or local authority region. For example, employment statistics in the tourism sector are generally aggregated for key tourist zones (such as Djerba-Zarzis in the Gulf of Gabès) whereas observed climate data refer to the closest meteorological station data (Sect. 5.3.3). In addition, marine observations (e.g., sea surface temperature, maximum significant swell height) are generally sparse and discontinuous in time and are integrated over a large spatial domain (Fig. 4.16).
- There tends to be less quantitative information about future trends and projections of socio-economic indicators particularly at a regional or local level. Even where projections are available, they tend to be for the short term (e.g., next year in tourism) rather than the longer time scales relevant to climate change.

These indicator themes and methodological challenges are developed further in Chap. 4. *Climate impact assessments* which presents examples of key impacts and vulnerabilities to climate of the present and recent past, by generic case study (urban, rural, coastal), and Chap. 5 *Integration of the climate impacts with future projections.*

References

Carter TR, Jones RN, Lu X, Bhadwal S, Conde C, Mearns LO, O'Neill BC, Rounsevell MDA, Zurek MB (2007) New assessment methods and the characterisation of future conditions. In: Parry ML, Canziani OF, Palutikof JP, van der Linden PJ, Hanson CE (eds) Climate change 2007: impacts, adaptation and vulnerability. Contribution of working group II to the fourth assessment report of the intergovernmental panel on climate change. Cambridge University Press, Cambridge, pp 133–171

Criado-Aldeanueva F, Del Rio Vera J, Garcia-Lafuente J (2008) Steric and mass-induced Mediterranean sea level trends from 14 years of altimetry data. Glob Planet Chang 60(3–4): 563–575. doi:10.1016/j.gloplacha.2007.07.003

EEA-JRC-WHO (2008) Impacts of Europe's changing climate – 2008 indicator-based assessment. European Environment Agency, Copenhagen, 246 pp

El-Askary H, Farouk R, Ichoku C, Kafatos M (2009) Transport of dust and anthropogenic aerosols across Alexandria, Egypt. Annales Geophysicea 27(7):2869–2879. doi:10.5194/angeo-27-2869-2009

Erhard M, Minnen JV, Voigt T (2002) Proposed set of climate change state and impact indicators in Europe. Technical report no. XX, Prepared for ETC. on Air and Climate Change, European Environment Agency, Bad Homburg, 37 pp

Esty DC, Levy M, Srebotnjak T, de Sherbinin A (2005) 2005 Environmental sustainability index: benchmarking national environmental stewardship. Yale Center for Environmental Law & Policy, New Haven, 414 pp

Füssel H-M (2007) Adaptation planning for climate change: concepts, assessment approaches, and key lessons. Sustain Sci 2:265–275. doi:10.1007/s11625-007-0032-y

Hák T, Moldan B, Dahl AL (eds) (2007) Sustainability indicators. A scientific assessment. Island Press, Washington, DC, 448 pp

Hammond A, Adriaanse A, Rodenburg E, Bryant D, Woodward R (1995) Environmental indicators: a systematic approach to measuring and reporting on environmental policy performance in the context of sustainable development. World Resources Institute, Washington, DC, p 43

JNF (2002) Report on drought damage in forests of the southern district. Jewish National Fund, Gilat, Israel (in Hebrew)

Jones RN (2001) An environmental risk assessment/management framework for climate change impact assessments. Nat Hazard 23(2–3):197–230. doi:10.1023/A:1011148019213

Kenny GJ, Warrick RA, Campbell BD, Sims GC, Camilleri M, Jamieson PD, Mitchell ND, McPherson HG, Salinger MJ (2000) Investigating climate change impacts and thresholds: an application of the CLIMPACTS integrated assessment model for New Zealand agriculture. Clim Chang 46(1–2):91–113. doi:10.1023/A:1005576910408

Meze-Hausken E (2008) On the (im-)possibilities of defining human climate thresholds. Clim Chang 89(3–4):299–324. doi:10.1007/s10584-007-9392-7

Moriondo M, Good P, Durao R, Bindi M, Giannakopoulos C, Corte-Real J (2006) Potential impact of climate change on fire risk in the Mediterranean area. Clim Res 31:85–95. doi:10.3354/cr031085

Niemeijer D, de Groot RS (2008) A conceptual framework for selecting environmental indicator sets. Ecol Indic 8(1):14–25. doi:10.1016/j.ecolind.2006.11.012

OECD (1993) OECD core set of indicators for environmental performance reviews: a synthesis report by the group on the state of the environment no. 83. Organisation for Economic Co-Operation and Development, Paris, 39 pp

OECD (2008) Key environmental indicators. Organisation for Economic Co-Operation and Development Environment Directorate, Paris, 36 pp

Peterson TC, Manton MJ (2008) Monitoring changes in climate extremes: a tale of international collaboration. Bull Am Meteorol Soc 89(9):1266–1271. doi:10.1175/2008BAMS2501.1

Risbey JS, Kandlikar M (2007) Expressions of likelihood and confidence in the IPCC uncertainty assessment process. Clim Chang 85(1–2):19–31. doi:10.1007/s10584-007-9315-7

Smith JB, Schellnhuber H-J, Mirza MMQ, Fankhauser S, Leemans R, Erda L, Ogallo L, Pittock B, Richels R, Rosenzweig C, Safriel U, Tol RSJ, Weyant J, Yohe G et al (2001) Vulnerability to climate change and reasons for concern: a synthesis. In: McCarthy JJ, Canziani OF, Leary NA, Dokken DJ, White KS (eds) Climate change 2001: impacts, adaptation, and vulnerability. Contribution of working group II to the third assessment report of the intergovernmental panel on climate change. Cambridge University Press, Cambridge, pp 913–967

Chapter 4
Climate Impact Assessments

Debbie Hemming, Maureen D. Agnew, Clare M. Goodess,
Christos Giannakopoulos, Skander Ben Salem, Marco Bindi,
Mohamed Nejmeddine Bradai, Letizia Congedi, Camilla Dibari,
Hesham El-Askary, Mutasem El-Fadel, Mohamed El-Raey, Roberto Ferrise,
José M. Grünzweig, Ali Harzallah, Abdallah Hattour, Maria Hatzaki,
Dina Kanas, Piero Lionello, Mark McCarthy, César Mösso Aranda,
Theib Oweis, Joan Pau Sierra, Basil Psiloglou, Marco Reale, Agustín
Sánchez-Arcilla, Mohamed Senouci, and Annalisa Tanzarella

Abstract This chapter highlights key climate impacts, hazards and vulnerabilities and associated indicators that have been used to assess current (recent) climate impacts at each of the case-study sites. The aim is to illustrate some of the wide range of information available from individual case studies and highlight common themes that are evident across multiple case-study locations. This is used to demonstrate linkages and sensitivities between the specific climate impacts of relevance for each case-study type (urban, rural and coastal) and the key climate hazards and biogeophysical and social vulnerabilities representing the underlying drivers and site conditions. For some

---

D. Hemming (✉) • M. McCarthy
Climate Impacts Analysis and Impacts Model Development Teams,
Met Office Hadley Centre, Exeter, Devon, UK
e-mail: debbie.hemming@metoffice.gov.uk

M.D. Agnew • C.M. Goodess
Climatic Research Unit, School of Environmental Sciences,
University of East Anglia, UEA, Norwich, UK

C. Giannakopoulos • M. Hatzaki • B. Psiloglou
Institute for Environmental Research and Sustainable Development,
National Observatory of Athens, Athens, Greece

M. Bindi • R. Ferrise
Department of Plant, Soil and Environmental Science (DIPSA),
University of Florence, Florence, Italy

Institute for Biometeorology, National Research Council of Italy,
(CNR-IBIMET) Florence, Italy

S. Ben Salem • M.N. Bradai • A. Harzallah • A. Hattour
Institut National des Sciences et Technologies de la Mer – INSTM, Salammbô, Tunisia

L. Congedi
Centro Euro Mediterraneo per i Cambiamenti Climatici, CMCC, Lecce, Italy

C. Dibari
Department of Plant, Soil and Environmental Science (DIPSA),
University of Florence, Florence, Italy

impacts, there are clear, direct links with climate events, such as heat stress and flooding, while for others, such as energy supply and demand, the causal relationships are more indirect, via a cascade of climate, social and economic influences. Water availability and extreme temperatures are common drivers of current climate impacts across all case studies, including, for example, freshwater supply and heat stress for urban populations; irrigation capacity and growing season length for agricultural regions; and saltwater intrusion of aquifers and tourist visitor numbers at coastal locations. At some individual case-study locations, specific impacts, hazards and/or vulnerabilities are observed, such as peri-urban fires in Greater Athens, infrastructure vulnerability to coastal flooding in Alexandria, groundwater levels in Tel Hadya and vector-borne diseases in the Gulf of Oran. Throughout this chapter, evidence of current climate impacts, hazards and vulnerabilities from each of the case studies is detailed and assessed relative to other case studies. This provides a foundation for considering the wider perspective of the Mediterranean region as a whole, and for providing a context from which to assess consequences of future climate projections and consider suitable adaptation options.

Keywords Climate change • Mediterranean • Impacts • Climate variability • Vulnerability

---

H. El-Askary
School of Earth and Environmental Sciences, Schmid College of Science and Technology, Chapman University, Orange, CA, USA

Department of Environmental Sciences, Faculty of Science,
Alexandria University, Alexandria, Egypt

M. El-Fadel
Faculty of Engineering and Architecture, American University of Beirut, Beirut, Lebanon

M. El-Raey
University of Alexandria, Alexandria, Egypt

J.M. Grünzweig • D. Kanas
Institute of Plant Sciences and Genetics in Agriculture, The Robert H. Smith
Faculty of Agriculture, Food and Environment, The Hebrew University of Jerusalem, Rehovot, Israel

C. Mösso Aranda • J. Pau Sierra • A. Sánchez-Arcilla
Laboratori d'Enginyeria Marìtima (LIM/UPC), Universitat Politécnica de Catalunya, Barcelona, Spain

T. Oweis
International Center for Agricultural Research in the Dry Areas, ICARDA, Aleppo, Syria

M. Reale • A. Tanzarella
Department of Material Science, University of Salento, Lecce, Italy

M. Senouci
Association Recherche Climat et Environnement – ARCE, Oran, Algeria

P. Lionello
Department of Material Science, University of Salento, Lecce, Italy

Centro Euro Mediterraneo per i Cambiamenti Climatici, CMCC, Lecce, Italy

4.1 Introduction

The CCIF framework (Sect. 1.3) has been used to identify key climate-related impacts, hazards and vulnerabilities for each of the CIRCE case-study sites (outlined in Chap. 1, with terms defined in Box 1.1), in association with regional stakeholders (see Chap. 2), and to establish suitable indicators and datasets for assessing changes in these (Chap. 3 and Appendix 2). In this chapter, we collate key highlights of current climate impacts, hazards and vulnerabilities systematically for each of the three case-study types – urban, rural and coastal. Understanding these current climate hazards and vulnerabilities, and their associated impacts, provides the context for considering future changes and assessing appropriate adaptation measures, which are discussed in Chaps. 5 and 6 respectively.

4.2 Urban

Many Mediterranean coastline cities, especially those in the east and the south of the region, currently face multiple challenges that can be attributed, at least in part, to characteristic environmental and socio-economic conditions experienced across the wider Mediterranean region. For example, features of the Mediterranean climate (see Volumes 1 and 2) such as heatwaves and flash floods, human health effects of air and water pollution, saltwater intrusion of coastal aquifers, and an inadequacy of energy and water resources exert stresses on the population health, infrastructure and functioning of urban areas.

Urbanization itself greatly affects surface characteristics which have feedback effects on the local and regional environment. These can result in distinct urban climates that differ substantially from those of surrounding rural environments. The most apparent consequence of this is the Urban Heat Island (UHI) which can contribute to higher temperatures and changes in other climatic characteristics, i.e. humidity, within cities compared to their rural surroundings (see Box 4.1). This occurs through the capacity for the built environment to store heat during the day and release it at night, along with the direct release of heat through human activity (for example heating or cooling of buildings, traffic, and human metabolism). The UHI is sensitive to the ambient climate, and it interacts non-linearly with the local atmosphere to also modify these climate conditions.

The urban case studies assessed in CIRCE: Athens, Greece; Beirut, Lebanon; and Alexandria, Egypt (see Sect. 1.4.2 for brief descriptions of these urban areas), experience a range of climate-related impacts as a result of their varying interactions with and exposure to different climate-related hazards and vulnerabilities. Figure 4.2 summarizes the key impacts, hazards and vulnerabilities identified and assessed for these three urban areas during the CIRCE project. These are discussed in detail in the following sections.

Box 4.1 Urban Heat Island Modeling for CIRCE Urban Case Studies

The Urban Heat Island (UHI) for the three Mediterranean cities, Athens, Alexandria and Beirut, has been explored by examining the frequency of summer extreme temperatures, accounting for the cumulative impacts of urban heat islands, climate change, and anthropogenic heat emissions in a series of model sensitivity studies. The UK Met Office urban surface scheme coupled to the regional climate model cannot account for regional variations in urban morphology (building types, configuration and properties), or for small scale (less than 25 km) variations in topography and other important micro-climate effects at the scale of the urban case studies, but it is able to simulate the important characteristics of a Mediterranean urban heat island and compares favorably to observations for the case study of Athens. The primary purpose of the model is not to quantify the heat island for the case studies, but to objectively assess the relative sensitivity of regional climate to the combined forcing from greenhouse gases and urbanization.

The UK Met Office regional and global climate models (McCarthy 2009 and McCarthy et al. 2010 respectively) both suggest that urban heat islands in the Eastern Mediterranean increase in magnitude in response to greenhouse gas induced climate change. For a simple comparison of hazards in the three case study cities, threshold values of 25 and 40°C were applied for Tn and Tx, respectively. For Athens, the modeled urban heat island increases the number of hot nights by approximately 3 weeks, which is a similar impact to simulated climate change by the 2050s (2041–2060). However, the cumulative effect of climate change and the urban heat island results in 75% of summer nights in the 2050s exceeding the originally defined threshold. While the urban heat island does not have a significant impact on the frequency of hot days for the present day period in Athens, it does result in an additional 10 hot days per year for the 2050s scenario. The frequency of hot nights is clearly sensitive to the cumulative impact of climate change and urban heat island in Alexandria and Beirut. In urban areas of Beirut, 'Tn exceeding 25°C' and 'Tx exceeding 40°C' are 1-in-20 year events for the present day, but become multi-annual events through a combination of climate change and the urban heat island by the 2050s (Fig. 4.1).

The main emphasis of these results is on the sensitivity to urban-induced climate change and they do not represent a robust prediction of future climate change effects in any one of these locations. They do, however, demonstrate that if urban heat islands are ignored the frequency of extreme temperature events in the future will be grossly underestimated. For example, hot nights for the non-urban surface in Athens rise from four events per year in 1971–1990 to 20 events per year in 2041–2060, but for the urban surface

(continued)

Box 4.1 (continued)

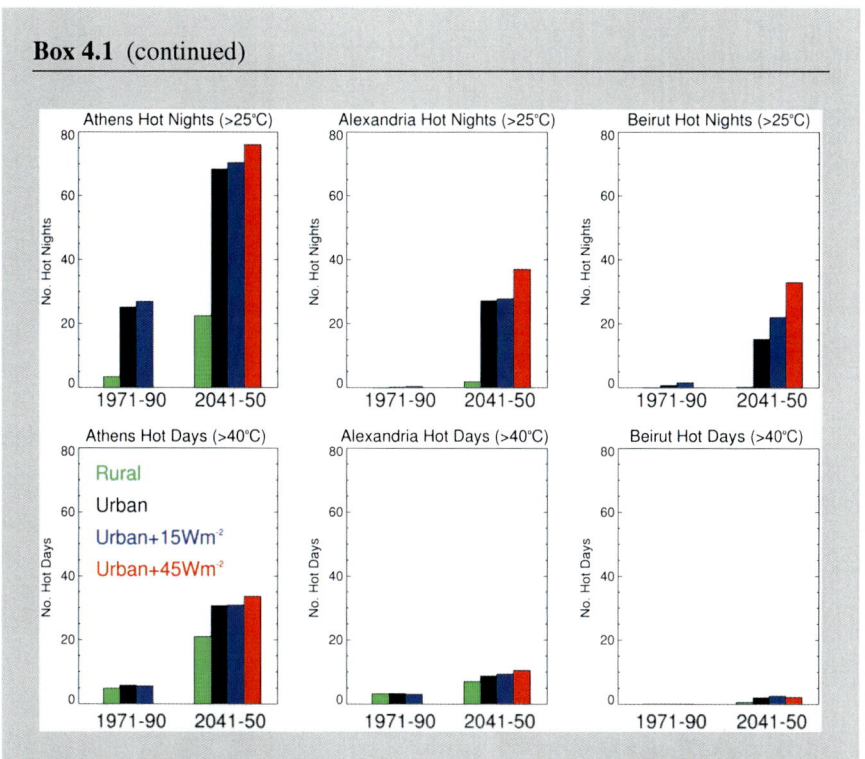

Fig. 4.1 Frequency of hot nights and days for a present (1971–1990) and a future period (2041–2060) for rural and urban areas, including an additional driver of urban climate change from local anthropogenic heat release (Urban +15 Wm^2, Urban +45 Wm^2)

the rise is from 25 to 75 events per year. Urbanization trends in the region could significantly exacerbate the heat-related impacts of climate change.

4.2.1 Climate Related Impacts

Many of the impacts and risks faced by cities across the Mediterranean region are in part related to their prevailing climatic conditions. From the Alexandria, Athens and Beirut CIRCE case studies, five key categories of impacts have been identified, after consultation with stakeholders, as being particularly sensitive to weather and climate, these are:

- Freshwater resources – quality and availability (Alexandria, Athens, Beirut)
- Human health (Alexandria, Athens, Beirut)
- Energy – consumption and production (Alexandria, Athens)
- Peri-urban fires (Athens)
- Infrastructure – coastal (Alexandria)

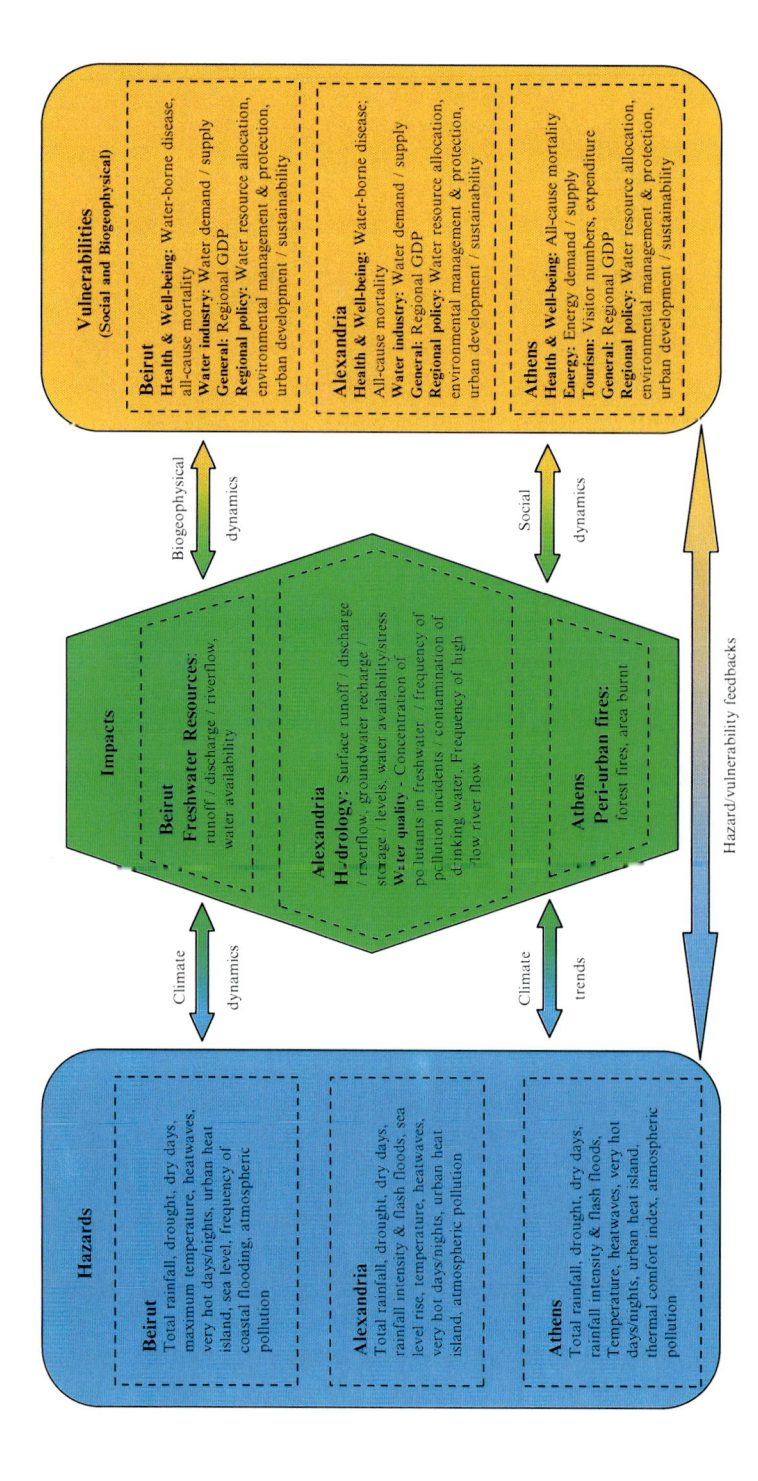

Fig. 4.2 Summary of key climate-related impacts at the three CIRCE urban case study locations together with the associated hazards and vulnerabilities (both social and biogeophysical). *Arrows* represent interactions between these categories via climate trends and dynamics and biogeophysical and social dynamics

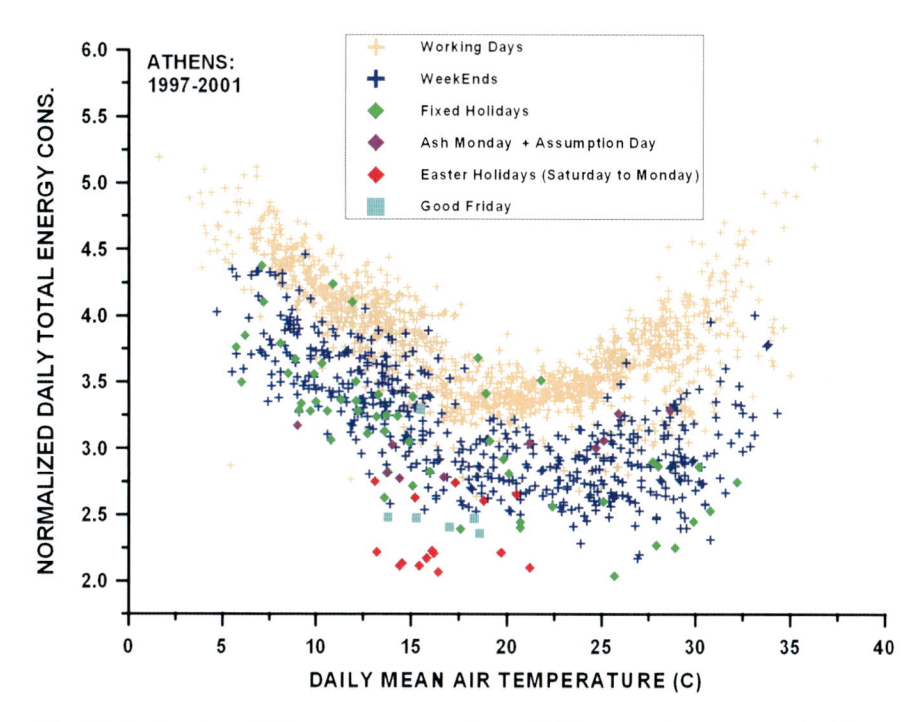

Fig. 4.3 Scatter plot of daily energy consumption and daily mean air temperature in Athens, 1997–2001

In some of the impact sectors, for example energy consumption (Fig. 4.3), the relationships between climate and impact are well studied and understood, so a quantitative approach has been followed with impact model construction and validation based on observed data. In these cases, it is possible to identify quantitative system thresholds (e.g., Table 4.1 for Athens, and for key health impacts in Beirut – see below for further discussion on these). In other sectors and in urban sites where observations are not available (e.g., health impacts in Alexandria), a more qualitative approach has been followed based on discussions and consultations with local stakeholders.

Given the nature of precipitation variability across the Mediterranean region (see Part I, Chap. 2) and the demand for freshwater to support urban populations, the availability and quality of freshwater resources exert major limitations on the design and function of urban systems, and their surrounding environments. These can manifest in a wide range of impacts depending on local characteristics and vulnerabilities. For example, in Athens the energy industry is particularly susceptible to changes in the availability and variability of freshwater, whereas in the densely populated urban area of Beirut water resources are largely utilized for human consumption with water tankers routinely required during dry periods to supply adequate quantities of freshwater to the rapidly increasing urban population. Water quality in Beirut is currently at risk from salinization of aquifers, where measurements show that the average salinity in freshwater aquifers has been continuously increasing over the past decades, starting with 340 mg/L in 1970, to 1,200 mg/L

Table 4.1 Summary table of biogeophysical and social vulnerability indicators for current climate in Athens, including identified system thresholds

Key impacts indicator	Climate hazard indicators	Vulnerability indicators	System thresholds	Current impacts
Fire weather index	Maximum temperature Relative humidity Wind Precipitation	Fire danger risk	Threshold for the daily fire occurrence is FWI >15	Fire risk increases significantly with FWI >30
All-cause daily death rate	Very hot days	Human health	Very hot days threshold = 34°C	Heat-related deaths increase exponentially >34°C
Electricity consumption	Temperature Heating degree days Cooling degree days	Energy consumption	Minimum energy consumption occurs around 22°C	Energy demand peaks in winter and summer and during weekdays
Air pollution	Temperature	Maximum daily temperature	Threshold for ozone exceedance days is T >18°C	The annual number of ozone exceedance days ranges from 30 to 90

in 1979, to about 4,200 mg/L in 1985 (Khair et al. 1992), and most recently, a study performed in 2004 revealed a salinity value above 5,000 mg/L at some locations (Saadeh 2008). In Alexandria, freshwater resources are required for domestic, industrial and tourist functions and the current freshwater resources are also at risk from salinization of the coastal aquifer and pollution from industry.

High variability in river flows, freshwater supply and extreme temperatures all pose challenges for energy production across the Mediterranean region. This is further compounded in urban areas by pressures of population growth and intra-annual variations in energy consumption from competing industrial, domestic, agricultural, and tourist activities. In Athens, for example, energy demand peaks in winter (due to heating), summer (due to cooling) and during the weekdays (due to work activities), and it has been noted that minimum energy consumption occurs at daily temperatures around 22°C (Table 4.1 and Fig. 4.3).

Human health is also a key impact category identified in all CIRCE urban case studies. Extreme temperatures and flooding events are features of the Mediterranean region and they are associated with widespread health impacts, including heat stress and exposure to diseases. Socio-economic and environmental pressures on freshwater availability, saltwater intrusion into aquifers and contamination of water supplies by poorly-regulated industrial activities and water-borne diseases, can result in poor water quality that is particularly important for human health in highly populated areas. Alexandria, for example, suffers from severe water pollution, largely from industrial activities and a lack of environmental monitoring and regulation.

In Beirut, heat stress and water pollution are key climate-related impacts on human health. The only empirical temperature-mortality relationship (derived for a short time period 1997–1999) for Beirut city has been established by El-Zein et al. (2004) and El-Zein and Tewtel-Salem (2005), and several studies have identified relationships between food and water borne diseases and climate variables (e.g., Singh et al. 2001; Hashizume et al. 2007; Chou et al. 2010; Checkley et al. 2000). Relationships from these past studies together with analyses during CIRCE have been used to assess direct health impacts in Beirut.

In other urban areas, interactions between localized topographic characteristics, urban activities and specific climatic events, result in episodes of severe air pollution that can have a dramatic impact on the health of the local population. This is the case in Athens, where surrounding mountains and urban activities favor the formation of severe air pollution episodes during periods of anticyclonic atmospheric circulation. Furthermore, in Athens the concentration of one of these health related air pollutants, low-level ozone, has been shown to be associated with temperature above a threshold of 18°C (Table 4.1). Athens already suffers from excess summer mortality as a result of extreme maximum temperatures (Fig. 3.8), with a very hot day threshold of 34°C above which heat-related deaths increase exponentially with temperature (Table 4.1). Such pollution events can act to compound these other health stresses.

Some climate-related urban impacts are not widespread across the Mediterranean region but rather relate to the particular features of the urban area. For example, coastal infrastructure in Alexandria is at risk from climate-related impacts of river and coastal flooding because of its low-lying deltaic location. Further, Athens is exposed to peri-urban fire risk because of its surrounding forest landscape, and it has been noted that fire occurrence (estimated using a Fire Weather Index (FWI) that is related to maximum temperature, relative humidity, wind and precipitation) starts to occur above a threshold FWI of 15, and increases significantly with FWI values above 30 (Table 4.1).

4.2.2 Climate Hazards

Key features of the Mediterranean climate pose hazards for the environment and people of this region, and, depending on the particular vulnerabilities, these may lead to the impacts outlined in Sect. 4.2.1 above. Extreme events always happen from time to time but present and projected changes in their intensity and frequency can exacerbate their impacts in urban locations. For the Alexandria, Athens and Beirut CIRCE case studies, four key climate hazards have been identified:

- High temperatures and heatwaves (Alexandria, Athens, Beirut)
- Rainfall variability – droughts and floods (Alexandria, Athens, Beirut)
- Air pollution (Alexandria, Athens)
- Coastal flooding (Alexandria)

Extreme high temperatures, frequency of heatwaves, and droughts and floods are key climate hazards that are common to urban areas and in many places have been

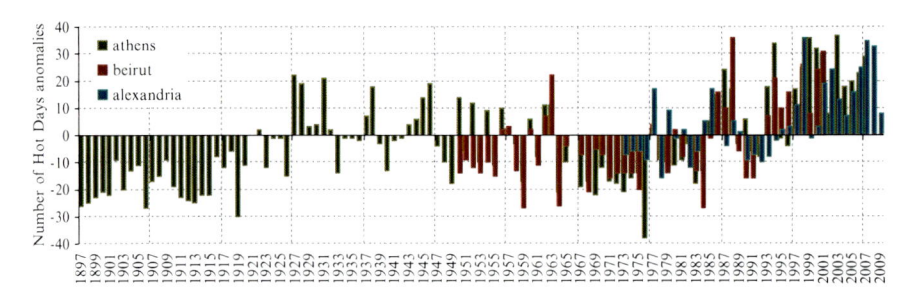

Fig. 4.4 Number of hot days (defined using the 90th percentile of daily Tx 1961–1990) shown as anomalies from the 1971 to 2000 average for Athens, Beirut and Alexandria

shown to be increasing across the Mediterranean region (see Part I, Sect. 2.3; Evans (2009)). These shape thermal comfort and cooling demands, and can have critical implications for surface water resources, peri-urban forestry, infrastructure, industry and human health. In coastal locations, sea-level rise and storm surges represent important hazards to coastal infrastructure and salinization of groundwater reserves. Flash floods are mainly a hazard for the northern Mediterranean cities, and saltwater intrusion a threat for low-lying and deltaic areas. In addition, for specific urban locations, such as Athens, air pollution events, that are associated with particular climatic and topographic conditions, pose significant hazards to human health. In Alexandria, a persistent aerosol cover exists over northern Egypt and the Greater Delta region; with dust aerosols dominating in spring, while urban/industrial emissions prevail during summer and autumn months. Thus aerosol optical depth is used as an indicator for the Alexandria case study (see Fig. 3.6).

An analysis of the long-term temperature records for Athens, Alexandria and Beirut demonstrates the more widespread tendency towards warmer average annual temperatures, with significant recent warming trends in summer maximum temperatures in Athens and Beirut, and in summer minimum temperatures in Alexandria. Since the 1970s, observed climate records from all three urban case study locations show a marked rise in maximum temperature (Tx) and the number of hot days per year (defined as days exceeding the 90th percentile of daily maximum temperature for 1961–1990), with a faster rate of increase apparent over the last two decades (Fig. 4.4). No statistically significant trend in total annual rainfall has been observed in any of the urban case study locations (Fig. 4.5).

4.2.3 Biogeophysical and Social Vulnerabilities

Vulnerability to climate change and variability is greatest for urban areas in the south and east of the region with their limited economic resources, rapid population growth and poor planning and regulation (Lionello et al. 2006). Levels of vulnerability and exposure also vary spatially within the urban area according to the chosen impact sector. More specifically, forest fire risk is primarily a concern in

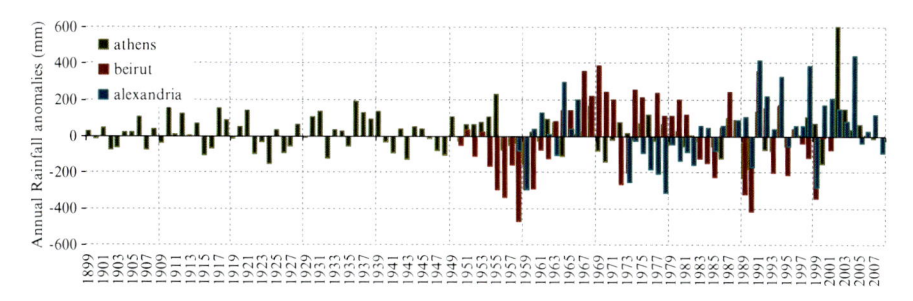

Fig. 4.5 Total annual rainfall shown as anomalies (mm) from the 1971–2000 average for Athens, Beirut and Alexandria

the peri-urban areas/suburbs of the city, whereas temperature-energy level fluctuations are an issue everywhere in the urban area, but there is a higher degree of exposure within the inner city where heat intensity is stronger, building density is greater and energy demand is larger. Heat stress is more likely to affect those living and working in the inner city (due to the urban heat island effect) than those living in the suburbs. In addition, elderly people and those with pre-existing cardio-respiratory problems are more at risk. Air pollution problems are also more likely in city areas close to emission sources (such as industrial estates and traffic dense zones and highways). The inner city zone is more at risk to air pollution from traffic emissions, while some suburban areas are located close to industrial activities (such as refineries on the west coast of Athens) and are also subjected to high levels of pollution. The suburbs are also vulnerable to high levels of secondary pollutants having undergone chemical transformation whilst being trans-ported according to the synoptic weather conditions.

For the three CIRCE urban areas, Alexandria, Athens and Beirut, key biogeophysical and social vulnerabilities have been identified (see Table 4.2). The greater administrative areas of Alexandria, Athens and Beirut represent a major part of the economic, social, cultural and administrative activities of their countries. Therefore, a severe heat wave in Athens for example, can paralyze all sectors and culminate in a substantial increase in energy demand (for air-conditioning) and a deleterious effect on human health. In Athens, urbanization (and high population density) constitutes an additional aggravating factor, and poor air quality further contributes to the increased health risks. Persistent high summer temperatures also significantly increase peri-urban forest fire risk (Giannakopoulos et al. 2009). Alexandria is marked by shortages in water resources that constitute a powerful constraint to continuing economic development. In addition, extreme climate events such as air pollution episodes or heat waves represent an additional challenge to the management of water resources, energy supply, and health services. Beirut, having nearly half of the population of Lebanon, is particularly vulnerable to potential climate change due to its coastal location, pre-existing poor quality of drinking water, water shortages, and deficient environmental management.

Although summer 2003 was exceptionally hot for the major part of the European continent, summer 2007 was exceptionally hot for many areas of south-eastern

Table 4.2 Key biogeophysical and social vulnerabilities identified for the CIRCE urban case study locations, Alexandria (Al), Athens (At) and Beirut (Be)

Biogeophysical vulnerabilities	Social vulnerabilities
Freshwater	**Freshwater**
Water availability/stress (Be)	Abstraction (Be, Al)
Soil moisture content (At)	Access to clean water per capita (Be)
Water for cooling plants (At)	Access to clean water per capita (Be)
Level of pollution (Al, Be)	Density of sewage treatment (Be)
Salt water intrusion (Al, Be)	Water use by sector and per capita (Be, Al)
	Drinking water contamination (Be)
Coastline flooding/retreat	**Infrastructure – industrial, residential**
Coastal inundation (Al)	% urban area in flood zone (Al)
Coastal erosion/retreat (Al)	GDP per capita (Al)
Agriculture/habitat	**Energy**
Intensity of agricultural systems (Be)	Population growth (Al, At)
Area of peri-urban forest (At)	Building density (Al, At)
Habitat suitability – vectors (Al)	Industrial output (Al)
	GDP per capita (Al, At)
	Energy facilities in flood zone (Al)
	Air conditioning proportion (At)
Air Pollution	**Tourism**
Vehicles per km² (Al, At)	Tourism facilities in flood zone (Al)
	% employed in tourism (At)
	Management/protection
	Expenditure on environmental protection/management (Be, Al, At)
	Water resource management (Be, Al)
	Investment in fire prevention/protection (At)
	Health
	Expenditure on health (Be, Al)
	Proportion of vulnerable population (<5, >65 years) (Be, Al, At)
	Access to health care (Al, At)
	% population with pre existing disease – cardiopulmonary (At)
	Demographic
	Population growth – density (Be, At)
	Housing density (Be, At)

Europe, the Balkan Peninsula and parts of Asia Minor with departures from the seasonal means exceeding 4°C in some areas with distinct periods of extremely hot weather. These unprecedented warm conditions were blamed for numerous excess deaths in several countries and serious problems in electricity supply in many regions. Greece experienced very likely the warmest summer of its instrumental history with record breaking temperatures being observed at a number of stations. The historical air temperature record of the National Observatory of Athens (NOA), extending back to the nineteenth century, was used in order to highlight the rarity of the event. Seasonal (June to August) temperature anomalies at NOA exceeded 3°C corresponding to more than three standard deviations with respect to the 1961–1990 baseline. The record value of 44.8°C was observed at NOA on 26 June 2007 (previous record 43°C in June 1916) during the first and most intense heat wave that affected the area. The analysis of temperature data from other less urbanized stations indicates that the urban heat effect in Athens contributed positively to the anomalies of the nocturnal temperatures. The abnormally hot summer of 2007 is perhaps not the proof but a strong indicator of what eastern Mediterranean summers could resemble in the future (Founda and Giannakopoulos 2009; see also Box 5.1).

Potential future changes for the urban case studies are discussed in Chap. 5 and adaptation options in Chap. 6.

4.3 Rural

Continued long-term climate variation (e.g., reduced rainfall, warmer average temperatures) and an increase in climate extremes (e.g., longer dry spells) could lead to an intensification of the constraints on rural activities across the Mediterranean region. The four rural case-study regions assessed in CIRCE (Apulia, Italy; Judean Foothills, Israel; Tel Hadya, Syria and Tuscany, Italy – see Sect. 1.4.3) are characterized by a range of features which are likely to render them particularly vulnerable to climate change.

Variation in rainfall amount and distribution, warmer temperatures, and the duration and intensity of extreme events (such as heat waves and protracted dry periods) are expected to have wide-reaching impacts on rural areas across the Mediterranean Basin (Giorgi and Lionello 2008; Lionello et al. 2006; Ulbrich et al. 2006; Luterbacher et al. 2006; Giorgi 2006; Lavorel et al. 1998) since they represent unique, fragile, and complex systems sustaining a multitude of function besides agricultural production (such as tourism, landscape, cultural identity, biodiversity, soil conservation and amenities).

Besides case-study specific climate indicators (see Chap. 3, e.g., Fig. 3.9), two core climate indicators were adopted in all the rural case studies: maximum temperature and precipitation anomalies calculated as the deviation from the long-term average. These indicators play a fundamental role in physical processes, ecosystem, biodiversity, surface water resources, agriculture, forestry, infrastructure, industry and human health, with consequences on the social and biogeophysical vulnerability of many human activities and natural systems.

There are large economic implications which may alter the characteristics of the workforce at a regional scale and the continuity of water supply to the general population. Moreover, in less developed areas, environmental change has immediate and direct effects on the health and well-being of households that depend on natural resources for their basic livelihood and could generate migratory fluxes that necessitate careful consideration and policy attention. Figure 4.6 summarizes the key

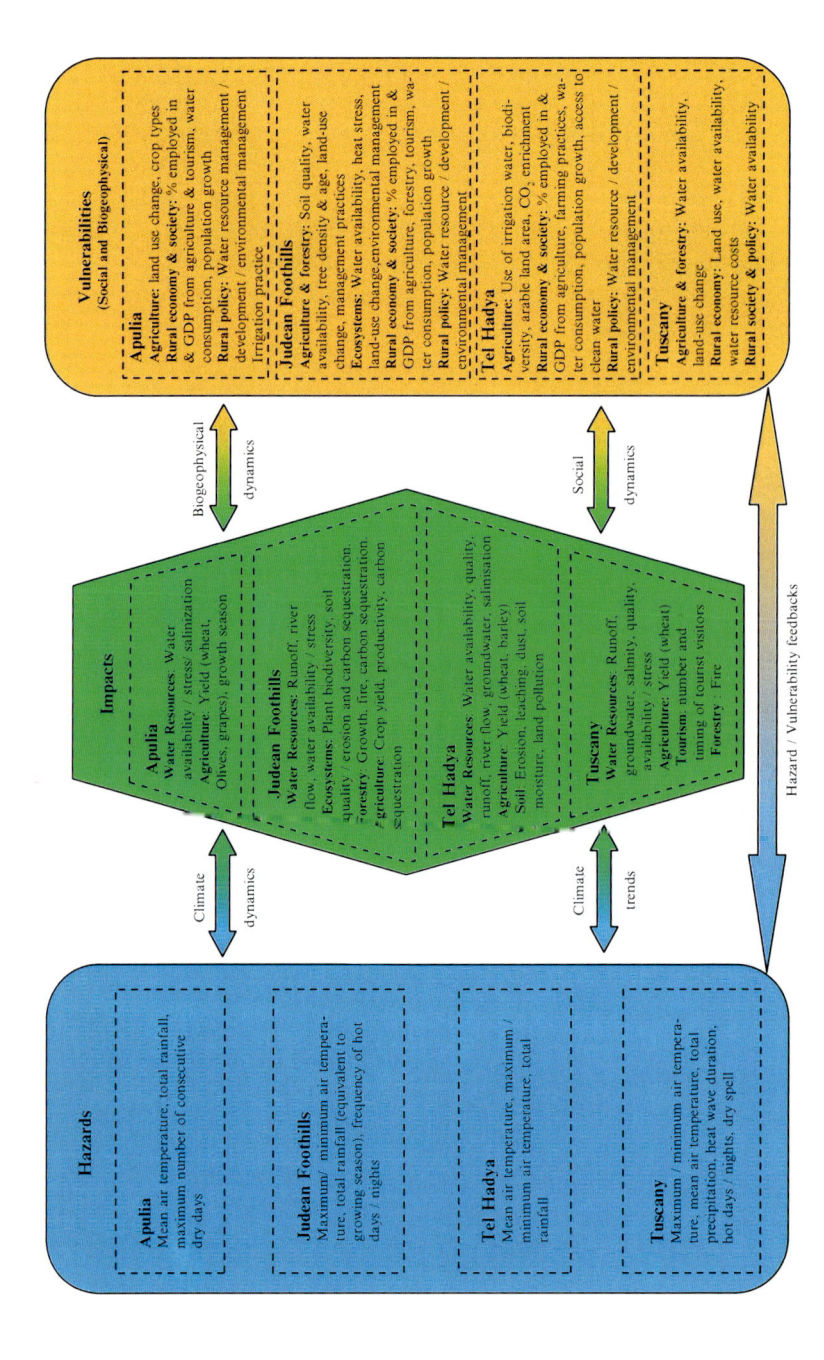

Fig. 4.6 Summary of key climate-related impacts (*green*) at the three CIRCE rural case study locations together with the associated hazards (*blue*) and vulnerabilities (*orange*) (both social and biogeophysical). *Arrows* represent interactions between these categories via climate trends and dynamics, and biogeophysical and social dynamics

impacts, hazards and vulnerabilities identified and assessed for the four CIRCE rural case studies. These are discussed further in the following sections, and expanded on in terms of future projections and potential adaptation options in Chaps. 5 and 6.

4.3.1 Climate Related Impacts

From the four rural CIRCE case studies of Apulia, Judean Foothills, Tel Hadya and Tuscany, six key categories of impacts have been identified as being particularly sensitive to weather and climate, these are:

- **Freshwater resources**
  - quality and availability (Apulia, Judean Foothills, Tel Hadya, Tuscany)
  - groundwater levels (Tel Hadya)
- **Agriculture**
  - crop yields, choice of crops, economic impacts on agriculture, carbon sequestration (Apulia, Judean Foothills, Tel Hadya, Tuscany)
- **Forestry**
  - productivity, carbon sequestration, forest fire (Judean Foothills, Tuscany)
- **Soil**
  - erosion/quality, carbon sequestration (Judean Foothills, Tel Hadya)
- **Ecosystems**
  - biodiversity, vegetation types (Judean Foothills, Apulia)
- **Tourism**
  - tourism (Apulia, Judean Foothills, Tuscany)

In all four case-study areas, changes in total annual and seasonal precipitation can lead to a serious reduction in water resource availability with unavoidable negative impacts on several economic sectors (including agriculture, tourism and industry) which compete for the same limited resource. For this reason, it is clear that an understanding of water resource variability, trends and projections, in the context of biogeophysical and social vulnerabilities is of paramount importance to many of the activities that take place in Mediterranean rural areas (Natali et al. 2009).

In particular, for all the rural case studies, changes in the availability and quality of freshwater resources, together with other climate variables, have key impacts on the rural economy via agricultural production of wheat (all four locations), barley (Tel Hadya), heavily irrigated vegetables including tomatoes (Apulia), and olives and grapes (Apulia and Tuscany), as well as influencing the choice of crops to grow and the carbon sequestration potential of the land (see Box 4.2). Moreover, water scarcity is one of the principal constraints of natural ecosystems and the services they provide in the Judean Foothills.

Box 4.2 Carbon Sequestration in the Judean Foothills

The region of the Judean Foothills comprises the transition zone between the northern sub-humid and the southern semi-arid Mediterranean climate zones of Israel. Many of the southern shrublands are degraded by different degrees of long-term overgrazing. These shrublands are probably depleted of organic matter, but have a large potential for carbon sequestration. This potential was exemplified by afforestation of a formerly overgrazed semi-arid shrubland. Thirty-five years after planting Aleppo pine (*Pinus halepensis*), the organic carbon content in the first 0.5 m of the soil profile increased by 75% relative to the native shrubland (Grünzweig et al. 2007). This corresponds to a storage rate of 50 gCm^{-2} year^{-1}, a rate rarely achieved by afforestation even in more moist climatic zones. At the ecosystem level, a mean carbon sequestration rate of 100 gCm^{-2} year^{-1} was obtained over 35 years (Grünzweig et al. 2007), with rates of the mature forest averaging 260 m^{-2} year^{-1}, as measured by eddy covariance (Rotenberg and Yakir 2010). Stable carbon isotope data showed that a large part of soil organic matter is included in mineral-associated fractions, and, thus, physically protected from decomposition (Grünzweig et al. 2007).

The Judean Foothills afforestation case study showed relatively large climate-change mitigation potential by semi-arid forests, in respect to carbon sequestration. However, it has lately been shown that planting trees on largely bare, overgrazed ground might positively feedback to global warming because of a decrease in short-wave albedo and a suppression of long-wave radiation as a result of low latent heat and high sensible heat fluxes (Rotenberg and Yakir 2010).

Carbon sequestration in the topsoil (0.1 m) was significantly affected by climate as shown by a comparison of two otherwise equivalent shrublands at the northern, sub-humid (540 mm mean annual precipitation) and the southern, semi-arid edge of the Judean Foothills (300 mm mean annual precipitation). Organic carbon stocks were reduced by 34% under shrubs and by 25% in herbaceous patches between shrubs, when moving from a sub-humid to semi-arid climate (Talmon et al. 2011). Because more organic carbon is stored under shrubs and because shrub cover decreases significantly from sub-humid to semi-arid conditions, the topsoil organic carbon stock at the ecosystem scale was lower by one third in the semi-arid shrubland. The decrease in organic carbon storage resulted from less input of new organic carbon (above ground biomass production decreased by 44% from north to south) and from continuing loss of older carbon (rate of litter decomposition decreased by only 29% from north to south). Therefore, relatively more organic matter decayed compared to the production of new biomass.

Water quality can also be affected by climate conditions, e.g., an increase in the frequency and intensity of droughts may cause saline intrusion in regions affected by water scarcity (as in the Tel Hadya, Apulia and the Judean Foothills case studies) with negative consequences on aquifers and on costs for water purification. Furthermore, in Apulia, unregulated use of wells and excess water extraction from aquifers exerts additional pressure on water resources.

The pressures and impacts of water scarcity often result in conflicts among users, and regional water resource managers face the dilemma of ensuring future sustainability of water resources while maintaining strategic agricultural, social and environmental targets.

Agriculture, one of the main sectors for the four rural case studies, may be greatly impacted by climate change. For example, high temperatures, coupled with water deficit, may negatively affect both natural and agriculture systems in terms of variations of plant phenology, growth rates, crop pests and diseases. Furthermore, as stated earlier by Mearns et al. (1996) and Long (1991), inter-annual rainfall variability is one of the major causes of variations in crop yields and yield quality (affecting photosynthesis, growth rate, nutrient acquisition and water status) and it may also affect crop management (e.g., time of sowing, fertilization, irrigation, harvest) through reductions in crop production and increases in management costs. Moreover, in a warmer and drier climate, additional crop irrigation water demand may also be requested leading to an increase in water abstraction with threatening consequences on fresh water quality and quantity due to the increasing competition for this limited resource.

Wine is an important product of Apulia which has produced in the last 25 years on average 9,800 million liters per year, representing approximately 15% of Italian national production. Apulia also produces about 2.15 million quart (1 L ~ 0.88 imperial quart) of olive oil (about 38% of Italian production) and 697,000 tons of wheat (about 8.5% of Italian production). The value of the 2008 year production was €270.1 million for wine, €376.92 million for olive oil and €324.18 million for wheat, representing about 1.2% of Apulia GDP for that year.

As well as agriculture, the rural case studies in CIRCE have highlighted that climate-related impacts are also of key importance to other aspects of rural landscapes and physical systems, most notably ecosystem health and function in the Judean Foothills, as indicated by plant species diversity and the diversity of vegetation types, forestry in both the Judean Foothills and Tuscany, from the perspective of forest productivity, forest fires, and carbon sequestration and soil in the Judean Foothills and Tel Hadya, in terms of erosion, quality and carbon sequestration. The diversity of natural vegetation types greatly decreases from the northern wetter part of the Judean Foothills to the southern drier part. While the northern part is characterized by various types of woodland and shrubland, the southern part is largely characterized by dwarf shrubland. This has consequences for regional diversity of plant species and for ecosystem services, such as prevention of soil erosion and grazing.

Previous research has shown that in natural shrubland, herbaceous species richness and the number of germinable seeds decrease with increasing aridity across the Judean Foothills (Holzapfel et al. 2006). Shrub cover, herbaceous biomass and total biomass decrease by up to 50% along a gradient ranging from 540 to 300 mm mean annual rainfall (Holzapfel et al. 2006; Talmon et al. 2011), which reduces the economic benefit from livestock grazing (Fleischer and Sternberg 2006). Organic carbon stocks in the topsoil and soil CO_2 efflux (a consequence of root and microbial respiration) have also been shown to decrease by about a quarter along the aridity gradient in this region (Talmon et al. 2011), whilst the total carbonate content of the topsoil increases almost seven-fold (Oren and Steinberger 2008). In forest plantations in this region, tree growth is significantly affected by water availability. A statistical relationship has been noted between El Niño events and above average precipitation/river flow (in the Jordan River) in the northern part of the area (Price et al. 1998), and these have been linked with enhanced growth of Aleppo Pine (Yakir et al. 1996). Tree mortality has been shown to increase greatly after two or more consecutive years of low rainfall (JNF 2002). In addition, late pine wood production starts earlier under dry than under wetter conditions, which has potential implications for tree growth and pest damage (Weinstein 1989). In the moisture-limited environment of the Judean Foothills, canopy photosynthesis and net CO_2 uptake tends to increase with increasing soil moisture availability (Afik 2009). Although annual rates of soil CO_2 efflux are less clearly related to total annual precipitation, they have been shown to increase linearly with water availability as expressed by soil water content or a water-availability index (Grünzweig et al. 2009).

Tourism is a major component of the rural economy in Apulia, the Judean Foothills and Tuscany and is highly sensitive to climate conditions (IPCC 2007). In fact, climate variability can provoke shifts in tourist flows and destinations, with large effects on the economic outcomes in regions where tourism-related activities represent a key source of income (e.g., in Tuscany the tourism sector accounts for 7–8% of regional GDP (Balestrieri 2005)). Tourism is an increasing resource for Apulia, where in the 10-year period from 1998 to 2007, tourist presence increased by about 68% from 1.36 million visitors to almost two million. Climate can directly influence tourism flows as, in the short term, favorable or unfavorable weather can by itself make a destination more or less attractive (e.g., more or less summer rain days at a seaside resort, and change in the number of hot days at inland natural parks and reserves). At the same time, climate can also indirectly affect in the longer term the quality of the amenities available at a given location (e.g., altered visual appeal of landscapes due to vegetation change; higher sea levels and beach erosion; water resource limitation).

4.3.2 Climate Hazards

Changes in precipitation and evaporation have key impacts which alter water resource availability in rural Mediterranean areas (Natali et al. 2009), and are often associated with compounding climate hazards, such as extreme temperatures, dust storms and forest fires. For the Apulia, Judean Foothills, Tel Hadya and Tuscany case studies, four key climate hazards have been identified:

- Rainfall variability – Annual rainfall, droughts and floods, rainfall intensity, consecutive dry days (Apulia, Judean Foothills, Tel Hadya, Tuscany)
- Temperature – growing season mean and variability, extreme hot temperatures, very hot days/nights, surface water evaporation (Apulia, Judean Foothills, Tel Hadya, Tuscany)
- Hot, dry winds, dust storms (Judean Foothills, Tel Hadya)
- Forest fire (Judean Foothills, Tuscany, Apulia)

In the agricultural sector, both the quantity and quality of production are strongly influenced by the amount and seasonal distribution of rainfall. In fact, crop production may be compromised by too much as well as too little rainfall during sensitive phenological stages (i.e., flowering and grain filling) (e.g., Tubiello et al. 2000). Rainfall patterns have extensive impacts on crop management systems, affecting for instance, the timing of sowing, fertilization, irrigation, harvest and so on. A reduction in precipitation across the Mediterranean Basin may raise water demands and intensify the sectoral (agricultural, civil and industrial – see Fig. 3.9) competition for water resources (consequently increasing the cost of water abstraction and likelihood of saline intrusion in coastal areas).

Temperature is a key indicator of climate change, and its maximum values (Tmax) represent one of the most sensitive variables to global warming in the Mediterranean Basin. A rise in Tmax may affect phenology, growth rates or water use of crops as well as the incidence of pests and diseases, particularly if critical thresholds are exceeded. Furthermore, a rise in Tmax may constrain many outdoor human activities, increase energy demands (e.g., for air conditioning) and impair human health (e.g., through heat stress).

There is an optimal temperature range for crops development that if exceeded plants become stressed or even they cannot cope at all and simply die. For example for wheat, both cold temperatures (<5°C) and hot temperatures (>30°C) at anthesis (fully open flower) can damage pollen formation, which in turn reduces grain set and yield (Dawson and Wardlaw 1989; Tashiro and Wardlaw 1990). The projected increase in temperature (see Sect. 5.2) may negatively impact crop yields due to the shorter duration of all crop developmental stages, allowing less time for capturing light, water and nutrient resources. High temperatures during reproductive development are particularly damaging.

For the CIRCE rural case study areas, changes in the frequency and/or intensity of rainfall and temperature, particularly via extreme weather events (such as flood, dry spells, heat waves and storms) form the main climate hazards. These also influence the occurrence of hot, dry winds and dust storms in the Judean Foothills and Tel Hadya.

Observed climate records covering the last few decades indicate that warming temperatures, highly variable precipitation amounts and distributions, and extreme events (such as heat waves and protracted dry periods) have been the main climate hazards that have impacted on the areas' agricultural productivity, natural ecosystems, forest fire risk, rural economy, tourism, and human health and well-being across all of the CIRCE rural case study areas.

Over the last few decades, warming trends were observed in all case study areas (see e.g., Fig. 3.5). An increasing and significant trend in annual temperature (with an accelerated rate in the last 30 years) is observed in all case-study areas. Temperature increases have been strongest during summer, particularly in Tuscany and the southern Judean Foothills, and at some of these sites, records of extreme temperatures have also shown warming. For example, across the Judean Foothills area, both the maximum and minimum mean annual temperature has increased by 0.2–0.4°C per decade since 1970, and the number of very hot days and nights has increased by ~3.5 and 6 days per decade respectively. The increase was maximal for summer temperatures in the southern part of the region (0.6 °C per decade). In Tuscany, the annual maximum temperature has increased by 0.3–0.4 °C per decade. This increasing trend has been observed across the whole region but the rate of increase is highest in inland areas and during the summer season. In the last 55 years, heat waves have steadily increased both in duration and frequency by 0.5–3 days per decade. Since 1951, the duration of heat waves in Tuscany has been extended by almost 10 days, and the longest events have occurred since 1990. In Apulia, for the whole period 1951–2005 there is little change in temperature, however, the second part of the record (1975–2005) shows a large increase (Fig. 4.7).

In all case-study areas, total annual rainfall showed high inter-annual variability over the common period (1971–2000), but no significant long-term trends. However, on a seasonal basis, significant decreasing trends in winter rainfall are evident for Tuscany (up to −13% per decade) and Apulia (around −10% per decade in January). A slower rate of decline in spring rainfall (up to −5% per decade) is also evident for Tuscany. For the coastal region of Apulia, sea-level rise and coastal storms are also key climate hazards; however, there is no clear evidence for long-term trends in marine storminess.

4.3.3 *Biogeophysical and Social Vulnerabilities*

Recent changes in biogeophysical and social dynamics, especially demography, tourism and agriculture, have led to increases in the vulnerabilities of many rural areas surrounding the Mediterranean, exerting additional pressures on natural resources in those regions. Table 4.3 shows the key biogeophysical and social vulnerabilities identified for the four CIRCE rural case study locations.

For all rural case study areas, it is clear that water resources and agricultural production are key biogeophysical and social vulnerabilities. Managing the competing water demands of household consumption, agriculture and tourism in areas where water supply from precipitation is highly variable is a key focus for the sustainable development of these areas.

Many rural Mediterranean areas are utilized for agricultural production, and are therefore particularly vulnerable to water availability, and other climatic changes.

4 Climate Impact Assessments

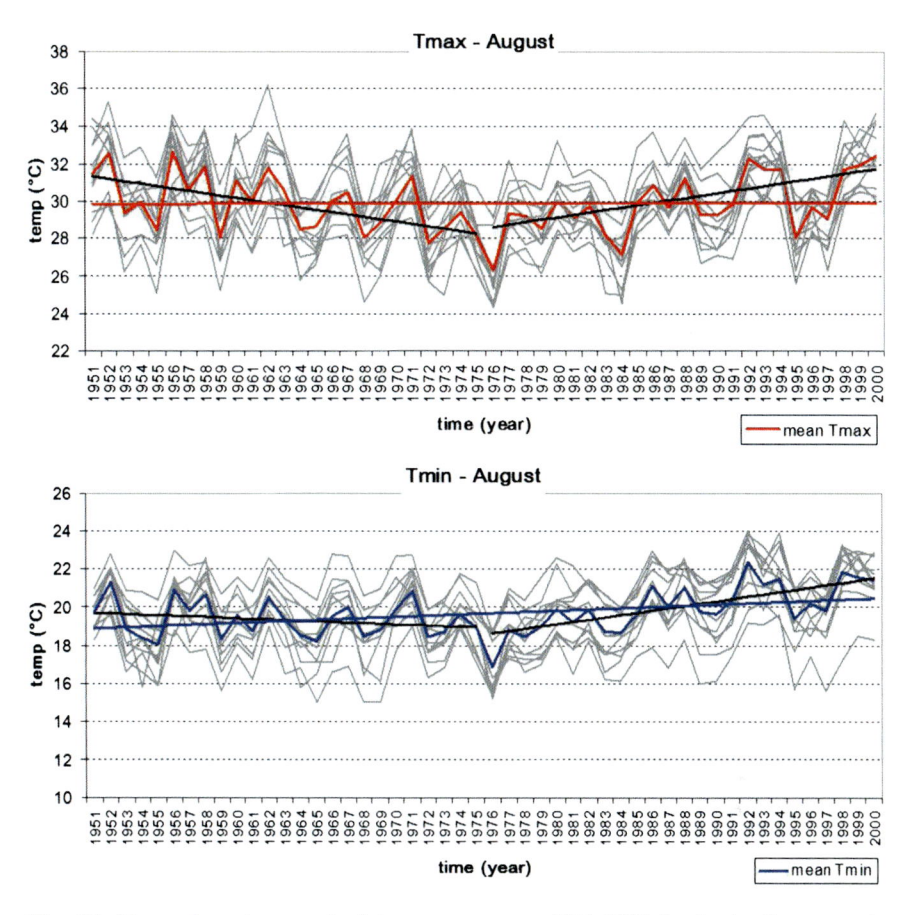

Fig. 4.7 Observed maximum and minimum temperature 1951–2000 for the Apulia case study site

In the Judean Foothills area, for example, 40% of the land area is used for agriculture and grain yields of rain-fed crops are sensitive to precipitation during the growing season (Fig. 4.8). Wheat grain is harvested between the end of May and the beginning of July, depending on the time of grain maturity. However, there is a threshold beyond which wheat is not harvested at all. If grain production during winter and spring is too low compared to the grain price and harvest costs, then the crop will be ploughed back in to the ground. During the 2007/2008 and 2008/2009 seasons, wheat was not harvested in the southern region of the Judean Foothills because rainfall during those years was 60–70 % of mean annual precipitation in the region. Even for irrigated crops across the Mediterranean, significant decreases in precipitation typically result in increases in the demand for already scarce water resources, which therefore feedback on socio-economic systems to increase regional vulnerabilities to climate change.

Table 4.3 Key biogeophysical and social vulnerabilities identified for the CIRCE rural case study locations, Apulia (Ap), Judean Foothills (JF), Tel Hadya (TH) and Tuscany (Tu)

Biogeophysical vulnerabilities	Social vulnerabilities
Freshwater	**Freshwater**
Abstraction – surface and groundwater, domestic, agriculture, industry, consumption (Tu, Ap, JF)	Abstraction/consumption – surface and groundwater by sector (Tu, TH, Ap, JF)
Salinization (Tu, TH, Ap)	Water resource management/protection (Tu, Ap, JF)
Water quality/pollution – oil spills, hazardous substances (TH)	Access to and use of fresh water per capita (TH, Ap, JF)
	Coastal protection (Ap)
Agriculture/forestry/ecosystems	**Agriculture/forestry/ecosystems**
Soil quality/erosion (TH, JF)	% employed in agriculture/forestry (TH, Ap, JF)
Biodiversity (TH, JF)	% contribution of agriculture/forestry to GDP (TH, Ap, JF)
Land use – Area of arable land/plantation forest/'natural' vegetation (Tu, JF)	Area of arable land (TH)
Pests and diseases (Tu, JF)	Intensity of farming/unsustainable farming techniques – livestock density, over cropping, over use of fertilizers (TH, JF)
Soil moisture content – agriculture, forestry (TH, JF, Ap)	
Water availability – agriculture (water resources per area) (Tu, TH, JF, Ap)	Conservation – protected land, number of environmental treaties, expenditure on nature conservation (JF)
Terrestrial pollution/leaching – oil spills, hazardous substances (TH)	
Water quality/pollution – oil spills, hazardous substances (TH)	
Critical species thresholds (JF)	
Tourism and rural economy	**Tourism and rural economy**
Traditional landscape – olives, vineyards etc. (Tu, Ap)	% employed in tourism (Ap, JF)
Land use (JF)	% contribution of tourism to GDP (Tu, Ap, JF)
Water availability (JF, Ap)	Economic growth (Ap)
	GDP per capita (TH, Ap, JF)
	Proportion of tourist facilities in coastal zone – vulnerable to flooding (Ap)
	Population density and growth (TH, Ap, JF)
	Use of supplemental irrigation – sustainability index (TH)
	Management/protection
	Expenditure on environmental protection (TH, Ap, JF)
	Water resource management (Tu, TH, Ap)

The rural environment of Tuscany, and in particular the agricultural sector, is considerably sensitive to climate variability (e.g., in temperature, precipitation and extreme events). In Tuscany, farmers drive cultivation towards higher incomes to cope with the ever increasing costs of production. Agriculture and related land use changes continuously in Tuscany in response to the Common Agricultural Policy

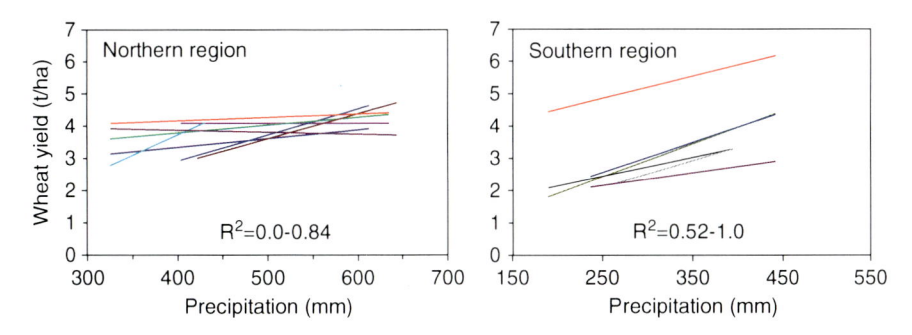

Fig. 4.8 Wheat grain yield (t/ha) of various rain-fed varieties at the farm level relative to total annual precipitation (mm) in two regions of the Judean Foothills

(CAP) and reform has arisen due to the effects of these changes (Bernetti et al. 2006). Tuscan farmers have thus more individual discretion to produce what the market demands in terms of 'quality', and are encouraged to apply agricultural practices and investments capable of guaranteeing a more stable and higher crop output. On the other hand, changes in climate variables may also be determinants in crop yield.

Long-term climate records for Tuscany reveal rising temperatures, and inter-annual variability in precipitation together with slightly lower total precipitation (see Sect. 4.3.2). Such changes are likely to have significant short- and medium-term effects on the availability of water resources, for both crop water requirements and irrigation use in the Mediterranean basin (de Wrachien et al. 2003). During periods of drought, an increasing amount of supplemental irrigation water is required to satisfy crop-water requirements for non-irrigated crops (e.g., wheat) and maintain baseline yields, thus further raising the demand for water. Adaptation and supplemental irrigation practices imply higher abstractions from surface and groundwater bodies, many of which are already in a critical condition in Tuscany. In addition, excessive water abstraction can cause salinization of groundwater, which particularly threatens agricultural activities, domestic water use and tourism along coastal areas. The volume of renewable water resources in Tuscany has decreased as a result of current climate hazards (Nuvoli and Matina 2002). A further example of water stress is the region of Apulia which is characterized by permanent water deficit (approximately −350 mm per year), with import of water from nearby regions and overexploitation and unregulated use of aquifers posing problems of sustainability and saline intrusion.

In regions of the Mediterranean that utilize groundwater for their supply of freshwater (see Part II), the quality and availability of water has been significantly affected by social factors, such as over abstraction, and biogeophysical factors, such as salt-water intrusion of coastal aquifers. In the Tel Hadya area, for example, groundwater resources are scarce, and abstraction of water for irrigation purposes has resulted in a substantial decline in groundwater levels (Fig. 4.9) and a number of wells drying up. Water quality is also an issue, and in some areas of the Tel Hadya region, the salinity of groundwater varies widely from an electrical conductivity

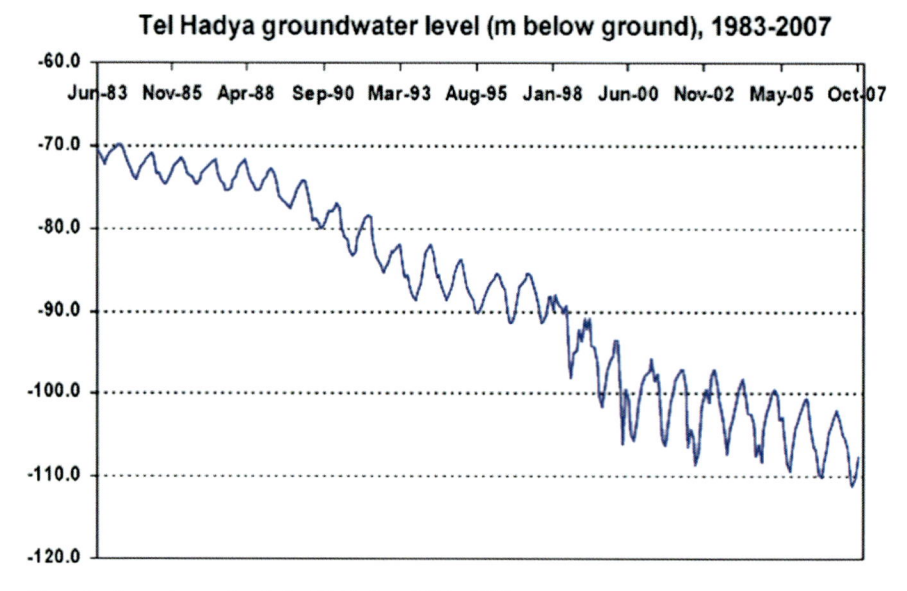

Tel Hadya groundwater level (m below ground), 1983-2007

Fig. 4.9 Groundwater levels in Tel Hadya, 1983–2007

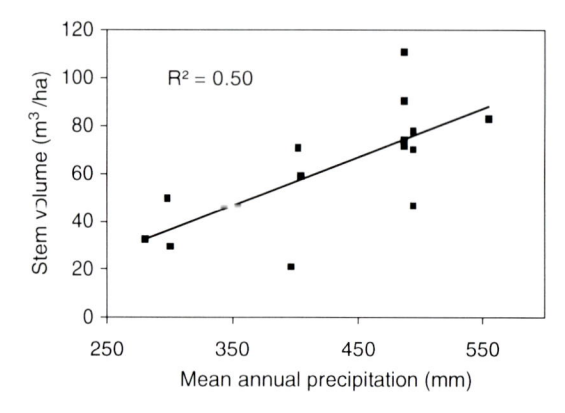

Fig. 4.10 Stem volume (m³/ha) of *Pinus halepensis* in planted forests (n = 14) along a rainfall gradient. Forests stands were aged 35–45 year with a density of 250–350 trees/ha. The linear regression is significant at p = 0.005 (Data source: JNF-KKL, Head Office, P.O. Box 7283, Jerusalem, Israel)

(EC) of 0.5–30 dS/m (salinity is measured as the EC expressed in units of deciSiemens (dS) per meter). Typical values of EC for irrigation water range from 1 to 4 dS/m, where 1 is excellent quality and 4 is low quality.

Studies of tree stem volume from forest plantations across a steep rainfall gradient in the Judean Foothills indicate a significant sensitivity of tree growth to water availability (in this case mean annual precipitation) in this region (Fig. 4.10). Tree stem volume is a measure of forest biomass at the stand level that integrates growth of individual trees and tree density of the stand. It is calculated from tree height and diameter, and is summed for all trees in a stand. Data are based on forest inventories conducted by the Jewish National Fund – Keren Hakayemet Leyisrael (JNF-KKL), the forest service of Israel. The results presented in Fig. 4.10 show the average stem volume for *P. halepensis* stands for different forest plantations along the rainfall gradient. These were selected within the dominant range of tree age and density to standardize as much as possible for other variables that affect stem volume in addition to rain amounts. Regression analysis of these data indicate that 50% of the variation in the average stem volume of *P. halepensis* stands across the region can be explained by the variability in mean annual precipitation ($R^2 = 0.5$, $P = 0.005$). Other controlling factors of production are site specific and might possibly be related to genetic differences among various provenances planted in the region. *P. halepensis* is the main tree species historically used for afforestation in the Judean Foothills. Mortality of forest trees is also related to water availability, as successive dry years can have impacts on tree survival. This additionally depends on tree species and stand characteristics, with dense stands of drought-sensitive species being mostly vulnerable to successive dry years.

Deforestation due to fire is of great concern for Tuscany, Apulia and the Judean Foothills since it can have widespread and damaging consequences through increased flood risk, soil erosion and loss of fertility (Court-Picon et al. 2004; Iliadis 2004). Forest fires also play a key role in determining the net carbon balance of forests (Kurz et al. 1995) and their greenhouse gas emissions (Flannigan et al. 2001). In Tuscany, although more than 90% of fires are started through arson, fires remain highly weather sensitive, particularly through the moisture content of the wood/fuel (affected by precipitation, relative humidity, air temperature and wind speed). Warmer air temperatures increase fuel dryness and reduce relative humidity resulting in a heightened fire risk particularly where rainfall is also declining (Howden et al. 1999). Physical and socio-economic vulnerabilities to fires and adaptation measures are also key factors influencing fire risk in specific locations and over time. For example, better management and control practices in Tuscany have been associated with decreases in the number of forest fires in the region (Fig. 3.3).

When changes in biogeophysical and social vulnerability are combined with trends in climate hazards, it is particularly difficult to understand the complex integration of processes that are attributed to resultant impacts. A good example of this is the differing land use trends that have been observed in Apulia and Tuscany over the last few decades. While both Apulia and Tuscany have experienced warming and drying over recent decades, in Apulia the low to moderate elevation topography and increasing population density have led to an intensification of agricultural practices and a successive replacement of existing natural vegetation with agricultural land. The area has experienced a spontaneous and progressive increase of irrigation practices (with a commensurate improvement in productivity) that has extended agricultural production from vegetables to tree crops, with intensive irrigation being used to grow table grapes and citrus, and emergency irrigation used to grow olives. In comparison, in Tuscany, the recent long-term trend in land use is towards less agricultural land. Due to the rural encroachment on urban places which peaked after the Second World War, Tuscan land use has notably changed over the last century. The large-scale abandonment of rural areas was one of the major driving forces behind changes in ecosystem function and dynamics (e.g., crop fields and grazing lands replaced by open unmanaged forests, shrublands or woodlands). The development of the agricultural sector has included structural transformation of farms, as well as land use and land cover changes, to meet market requirements in terms of economic efficiency. The Tuscan rural economy results from a complex interaction between biogeophysical and socio-economic factors operating at various levels and driving land use pattern modifications with implications for the multi-functionality of agriculture (Bernetti and Marinelli 2010). According to Corine Land Cover surveys (1990 and 2000), Tuscany has been affected by a notable reduction in agriculture land – heterogeneous agricultural areas experienced the greatest loss (48% of total) and urban areas experienced the largest gain (about 41%). This trend is also confirmed by Agnoletti (2007) who led a study on landscape changes from 1832 to 2006 on a hilly-mountain area of Tuscany. His study showed dramatic decreases in arable land, pastures and mixed cultivations with vines during this period. According to Agnoletti, the loss of arable lands and pastures and secondary succession occurring over this period explains the high expansion of forests from 2002. This is largely due to the inversion of demographic trends which appeared around 1921 and more rapidly after the Second World War, as Italy was transformed into an industrialized country. These trends are also reflected at the national level: during the last century in Italy 13 million ha of agricultural land was abandoned and forests increased from 3.650 million ha in 1906 to 10.5 million ha in 2006 (National Forest Inventory http://www.ifni.it) (Agnoletti et al. 2006).

Table 4.4 summarizes the key current climate impacts and biogeophysical and social vulnerabilities identified for the Tuscany case study. The dominant role of water availability (via precipitation and temperature) and extreme heat are clear from the observed climate sensitivities and thresholds of the three key impacts indicators in this region – annual wheat yields, tourist arrivals, Fire Weather Index. For example, between 1985 and 2005, the highest number of summer fire events in this region coincided with the highest summer FWI values (1985, 1993, 2003).

With large proportions of the Mediterranean region covered by natural ecosystems and agricultural production, many of these areas are sensitive to climatic changes. Recent trends in biogeophysical and social vulnerabilities, such as water pollution and population growth, have tended to increase the pressure on the ecosystem services that these regions provide, and indirectly impact on economic activities, such as agricultural production and tourism, and rural livelihoods. Sustainable management practices, particularly for water resources and land degradation, are required to maintain current activities in many rural areas across the Mediterranean. This will be particularly important with the additional stresses, in the form of increasing temperature and decreasing precipitation, which climate projections indicate may occur across the region (see Chap. 5), and adaptations to improve availability and limit losses of water will be key to the future sustainability of agricultural production, and associated rural economies and livelihoods. Such adaptation options are discussed in the context of the CIRCE case studies in Chap. 6.

4 Climate Impact Assessments

Table 4.4 Summary of biogeophysical and social vulnerability indicators for current climate in Tuscany

Key impacts indicator	Climate hazards indicator	Vulnerability indicators	System thresholds	Current impacts
Annual wheat yields (Mg/ha)	Annual precipitation anomalies Max temperature anomalies Heat wave duration Dry spell	*Water Availability*: Competition for water resources has intensified Water deficit can also result from mismanagement of water resources Higher water consumption by agriculture is observed during the warmest years (e.g., 2003) A reduction in precipitation leads to lower water storage and consequently less water availability *Water Resource Costs*: An increase in water resource demand and sector competition leads to a rise in the cost of water abstraction *Population and rural economy*: 40% of the Tuscan population lives in rural areas, covering 75 % of the total regional area, producing one third of the total regional added value, and contributing to 35 % of employment	**Qualitative thresholds identified**: *Cultivation costs*: when these are higher than grain production income, the crop may not be harvested or sowed *Precipitation*: When persistent autumn rainfall delays sowing until the end of February, wheat cultivation may not be remunerative even for late cultivars **Climate thresholds for wheat yield in a general Mediterranean context**: *Temperature*: <5 °C or >30 °C during reproductive development crop phases (Dawson and Wardlaw 1989; Tashiro and Wardlaw 1990)	A water deficit (~300 mm) was recorded in the lowlands of the region in 1999 and throughout the region in the extremely dry year 2003 The lowest mean annual yield is observed in 2003, notably the driest and warmest year in recent decades In 2008–2009, persistent rain during the sowing season caused a general reduction in areas of wheat cultivation with respect to the previous year (−39 % and −40 % for durum and bread wheat, respectively)
Tourist arrivals	recipitation anomalies Max temperature anomalies Heat wave duration	*Water Availability*: Water resources required for tourism services (i.e., swimming pools, drinking water, recreational water sports) can reduce water resource availability	**Qualitative thresholds identified**: Hotter summers are associated with fewer tourist arrivals, whereas higher temperatures in shoulder seasons seem to be beneficial to tourism (Morabito et al. 2004, 2005)	During the summer heatwave of 2003, 72.7 % of total hospital admissions in Florence were related to heat-related diseases compared to 27.3 % in 2002 (Morabito et al. 2005)

(continued)

Table 4.4 (continued)

Key impacts indicator	Climate hazards indicator	Vulnerability indicators	System thresholds	Current impacts
	Hot days/nights Dry spell	Summer peaks in tourism accentuate pressure on water resources during drought periods, endangering water resource sustainability A tourist consumes on average 300 L of water/day, and up to 880 L/day in luxury hotels (EEA 2003) *Land Use:* Changes in land use may alter the traditional landscape of the region and reduce tourist attractiveness (Morabito et al. 2005) *Other income for farmers:* In Tuscany, income from non-agricultural activities (2003–2008 average) represents ∼ €140 Million (i.e., 8% of total agriculture income)		According to stakeholder consultation and ISTAT analysis (IRPET 2009), in recent decades there has been an increase in tourist arrivals for agriturismi-farmhouses (especially with swimming pools) located in the Tuscan countryside
Fire Weather Index (FWI)	Annual precipitation anomalies Max temperature anomalies Heat wave duration Hot days/nights Dry spell	*Water Availability:* The main fire risk occurs in summer when droughts are more frequent, water resources more vulnerable and soil moisture scarcer. In these conditions, plants suffer water deficit and become more combustible *Land Use:* During the previous century, cultivated land in Tuscany has been abandoned and replaced with often unmanaged forests with a dense and very dry flammable understory	*No quantitative thresholds have been identified*	Between 1985 and 2005, the highest number of summer fire events coincided with the highest summer FWI values (1985, 1993, 2003)

4.4 Coastal

Coastal zones have some of the highest population densities in the Mediterranean area. They typically support high concentrations of uses and infrastructure, productive biological ecosystems and provide the vital link between terrestrial and aquatic systems. As such, coastal Mediterranean zones experience a wide range of climate-related impacts that are influenced by many interacting hazards and vulnerabilities.

The main hazards in Mediterranean coastal zones arise from the fact that they are in a dynamic and fragile equilibrium between terrestrial, marine, riverine and atmospheric systems. Changes in any of these systems can lead to a change in the status of the coastal zone which may not be compatible with its current uses and infrastructure.

Major social and biogeophysical vulnerabilities in Mediterranean coastal zones are characterized by the current high population density which results in coastal buildings and infrastructure being exposed to direct wave action, and a lack of space for these coastal systems to easily migrate to less vulnerable locations. The uniqueness of many coastal ecosystems in the Mediterranean also adds to the environmental vulnerability. For example, the low-lying deltaic areas of the Nile and Ebro rivers (two of the CIRCE coastal case-study locations) are subject to rapid rates of subsidence and are particularly vulnerable to changes in mean sea level or wave storminess. Changes in the mean sea level or wave storminess also affect the low-lying islands of Kerkennah, Djerba and Kneis in the Gulf of Gabès (also a CIRCE coastal case-study location).

Figure 4.11 summarizes the key impacts, hazards and vulnerabilities identified and assessed for the four CIRCE coastal case studies: Gulf of Gabès, Tunisia; Gulf of Oran, Algeria; Gulf of Valencia Catalan coast, Spain; and West Nile Delta, Egypt (see Sect. 1.4.4 for descriptions of the case-study locations). These are discussed further in the following sections.

4.4.1 Climate Related Impacts

From the four coastal CIRCE case studies of Gulf of Gabès (GA), Gulf of Oran (OR), Gulf of Valencia Catalan coast (VA) and West Nile Delta (WN), six key categories of impacts have been identified as being particularly vulnerable and sensitive to weather and climate, these are:

- **Coastal**
  - erosion (GA, OR, VA, WN)
  - flooding (GA, OR, VA, WN)
  - saline intrusion (GA, VA, WN)

- **Coastal ecosystems and fisheries**
  - biodiversity (GA, OR, VA, WN)
  - fish catch (GA, OR, VA, WN)

Hazards

Gulf of Gabes
Climate: Total rainfall, drought, dry days, maximum and min air temperature, very hot days, thermal comfort index, UHI, seasonal shift index, maximum wind speed, daily climate-tourism potential index
Marine: Annual maximum significant swell height, relative sea level, sea temperature

Gulf of Oran
Climate: Total rainfall & distribution, drought, air temperature, thermal comfort index, heat wave duration index, hot days, air pollution, (thunder) storms
Marine: Storm surges, sea level, sea surface temperature, ocean circulation

Valencia
Climate: Total rainfall, drought, maximum air temperature, heat wave duration, very hot days/nights, thermal comfort index
Marine: Storm surges (sea level), wave storms, significant wave height, sea surface temperature

West Nile Delta
Climate: Total rainfall, drought, dry days, maximum air temperature, very hot days/nights
Marine: Storm surges, significant wave height, relative sea level, sea temperature, ocean currents, salinity

Impacts

Gulf of Gabes
Coastal: Saline intrusion.
Coastal erosion - shoreline retreat, sediment supply, coastal flooding.
Ecosystems & Fisheries: Marine biodiversity -fish landed productivity.
Tourism: Attractiveness - beaches & weather
Water Resources: Water availability

Gulf of Oran
Coastal: Coastal erosion / degradation, coastal flooding
Ecosystems & Fisheries: Marine biodiversity
Water Resources: Water availability

Valencia
Coastal: Saline intrusion, coastal erosion, sediment supply, coastal flooding
Ecosystems & Fisheries: Marine Biodiversity, fish catch
Agriculture: Yield of key crops
Water Resources: Water availability

West Nile Delta
Coastal: Saline intrusion, coastal erosion & accretion, sediment supply, coastal flooding
Ecosystems & Fisheries: Fish catch, marine biodiversity
Agriculture: Yields
Water Resources: Water availability

Vulnerabilities
(Social and Biogeophysical)

All coastal case studies
Coastal: Coastal protection, coastal buffer, freshwater outflows (fluxes, sediment, nutrient)
Ecosystems & Fisheries: Fish stocks, water quality, freshwater outflows / nutrients, quotas
Water Resources: Water quality, vulnerable aquifers, salinisation, salt water wedges, consumption, river flow, abstractions
Coastal economy & society: Population growth & density, per capita GDP, % employed in & GDP from agriculture, fishing & tourism, tourist numbers & expenditure, land, people & facilities in flood zone, access to healthcare, suitable vector habitat.
Coastal policy: Water resource & environmental management, coastal protection, regional development

Case study specific
Gulf of Gabes
Ecosystems & Fisheries: Alien species, biodiversity, % employed & GDP from fishing
Agriculture: Soil erosion quality, salinisation, inundation.
Tourism : beach erosion, favorable weather conditions

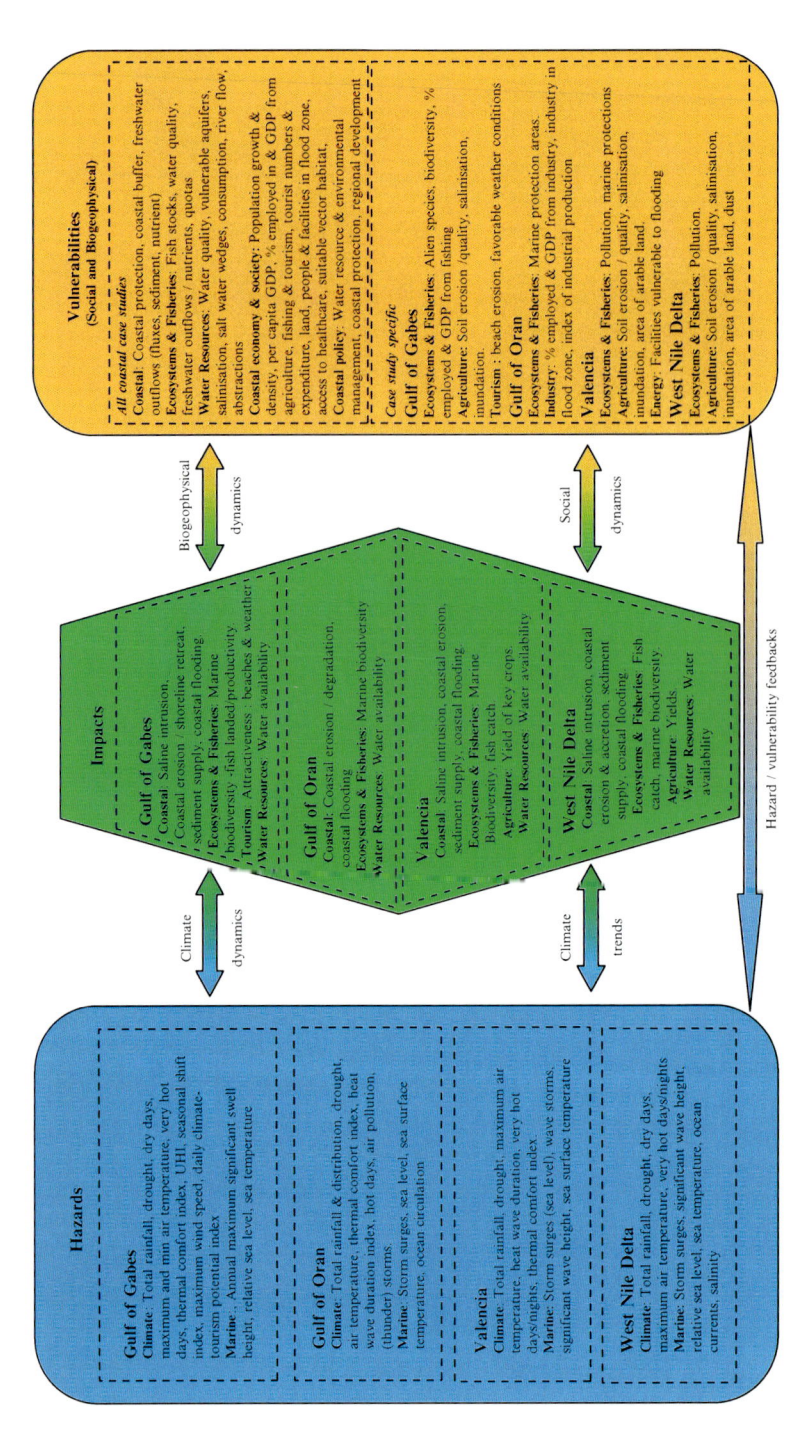

Gulf of Oran
Ecosystems & Fisheries: Marine protection areas.
Industry: % employed & GDP from industry, industry in flood zone, index of industrial production

Valencia
Ecosystems & Fisheries: Pollution, marine protections
Agriculture: Soil erosion / quality, salinisation, inundation, area of arable land.
Energy: Facilities vulnerable to flooding

West Nile Delta
Ecosystems & Fisheries: Pollution.
Agriculture: Soil erosion / quality, salinisation, inundation, area of arable land, dust

Fig. 4.11 Summary of key climate-related Impacts (*green*) at the four CIRCE coastal case study locations together with the associated hazards (*blue*) and vulnerabilities (*orange*) (both social and biogeophysical). Arrows represent interactions between these categories via climate trends and dynamics, biogeophysical and social dynamics, and hazard/vulnerability feedbacks

- **Freshwater resources**
  - availability (GA, OR, VA, WN)
- **Agriculture**
  - crop yields, economic impacts on agriculture (VA, OR, WN)
- **Urban infrastructure including energy, petrochemical**
  - building damage from surges (VA, OR)
- **Tourism**
  - numbers and visits to attractions (GA, VA)
- **Health**
  - heat and pollution related mortality (VA, OR)
  - vector-borne disease (OR)
  - hazardous storms (VA, OR)

Coastal processes of erosion and flooding, coastal biodiversity and fish catch, and freshwater availability have been identified as key sources of climate related impacts at all four case-study locations. In locations where agriculture is an important coastal industry (for the case studies this includes the Gulf of Valencia Catalan coast, Gulf of Oran and West Nile Delta), crop yields and other economic impacts on agriculture were also identified as key impacts. Likewise, where coastal aquifers are significant sources of freshwater (Gulf of Gabès, Gulf of Valencia Catalan coast and West Nile Delta) saline intrusion into these aquifers is also a key impact. Regionally, other particular impacts are important, for example, climate-related impacts on urban infrastructure (energy and petrochemical), tourism (beaches erosion and favorable/unfavorable weather conditions for tourism in the Gulf of Gabès) and human health (through heat, pollution and storms) have been identified as of key importance in the Gulf of Valencia-Catalan coast and the Gulf of Oran. For example, in the Gulf of Oran area, the sea is the direct discharge system for effluents from the town of Oran. Total waste-water discharge along the Gulf of Oran in 1995 and 2005 has been estimated for industrial plants (characterized by water consumption of more than 1,000 m^3/year) and domestic and collective discharges (see Table 4.5).

4.4.2 Climate Hazards

The main hazards for coastal zones arise from the dynamic and fragile equilibrium that exists between terrestrial, marine, riverine and atmospheric systems. A change in any one of these components, due to natural climatic variability or accelerated climate change will modify the characteristics of the coastal zone to a state which may not be compatible with present uses and infrastructure.

Table 4.5 Waste water discharge in the Bay of Oran (SOGREAH Engineering 1998)

	1995		2005	
	m³/day	%	m³/day	%
Domestic discharge	52,284	88	183,448	95
Industrial waste	6,933	12	9,584	5
Total discharge	59,217	100	193,032	100

For the Gulf of Gabès, Gulf of Oran, Gulf of Valencia Catalan coast and West Nile Delta case studies four main climate hazards have been identified (see also Sánchez-Arcilla et al. 2010):

- Sea level/storm surges – annual maximum significant swell height (GA) and significant wave height and duration (VA, OR, WN)
- Sea temperature (GA) and salinity, currents/circulation (VA, OR, WN)
- Air temperature – extreme air temperature, heatwave duration, thermal comfort, very hot nights (VA, OR, WN GA)
- Rainfall – Annual rainfall, droughts, consecutive dry days, and floods (VA, OR, WN)

Relative sea level, storm surges and wave action pose key hazards for the coastal case study locations because of their potential impacts on erosion, flooding, salinization of coastal aquifers and associated disruptions to ecosystems and human activities. There are no common records of sea level indicators at all of the case study sites, and the records that do exist show no clear consistency in recent sea level and/or storm trends. For example, sea level hindcast data for the Ebro Delta within the Gulf of Valencia Catalan coast indicate that the trend in sea level due to storm surges has decreased by about −0.54 cm/decade between 1958 and 2001 (Fig. 4.12), while a hindcast record of significant wave height for the nearby Catalan coasts indicates a slight increasing trend over the 1959–2001 period (Fig. 4.13). Annual maximum significant swell height records from three locations in the Gulf of Gabès for the period 1974–1993 show increasing trends (0.9–1.2 m per decade) (Fig. 4.14). The study period is short, however, so the trend may be strongly influenced by decadal variability. Finally, a record of the annual distribution of meteorological storms in the Gulf of Oran during 1950–2007 shows no significant trend (Fig. 4.15).

In contrast to the varying signals shown by sea level and storm records across the case-study locations, records of air and sea-surface temperature, and their extremes, show consistent warming trends in all locations. As examples, Fig. 4.16 shows the records of annual sea surface temperature (minimum, mean and maximum – 1969–2008) and mean air temperature (1905–2008) for the Gulf of Valencia Catalan coast study area, and Fig. 4.17 shows the records of annual mean maximum temperature anomalies (from the 1971–2000 baseline) and the annual number of hot days (>90th percentile of the 1971–2000 period) for both the Gulf of Oran and West Nile Delta study locations. The warming trend of the annual mean temperature at the Gabès observation station (Gulf of Gabès) for the period 1973–2008 (Fig. 4.17), is 0.46°C/

4 Climate Impact Assessments

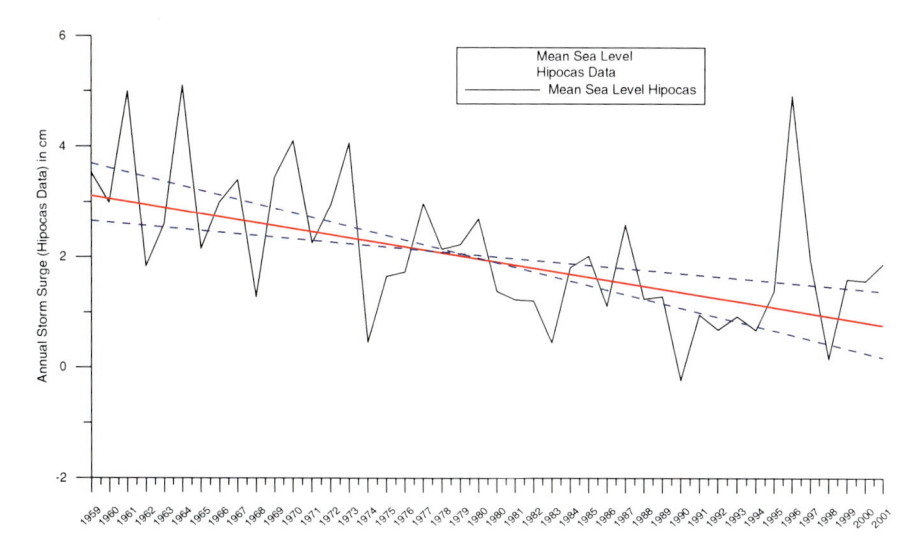

Fig. 4.12 Annual mean sea level (cm) at the Ebro Delta, 1958–2001 (HIPOCAS hindcast model data) together with the slope estimate (*red line*) with the 95% confidence interval (*blue dashed lines*) (Modified from Mosso et al. 2009)

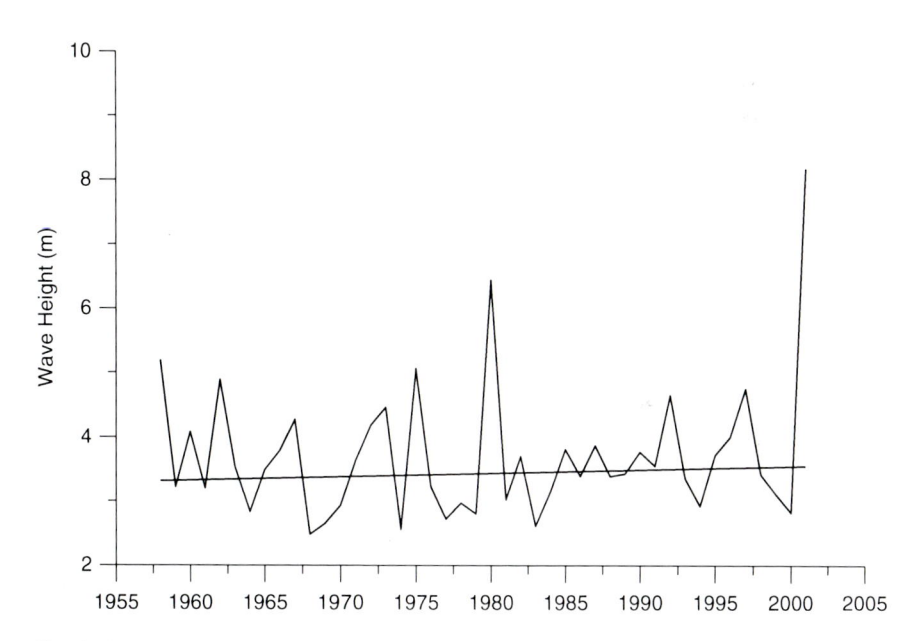

Fig. 4.13 Yearly maximum hindcast significant wave height values (m) for the Catalan coast (1959–2001)

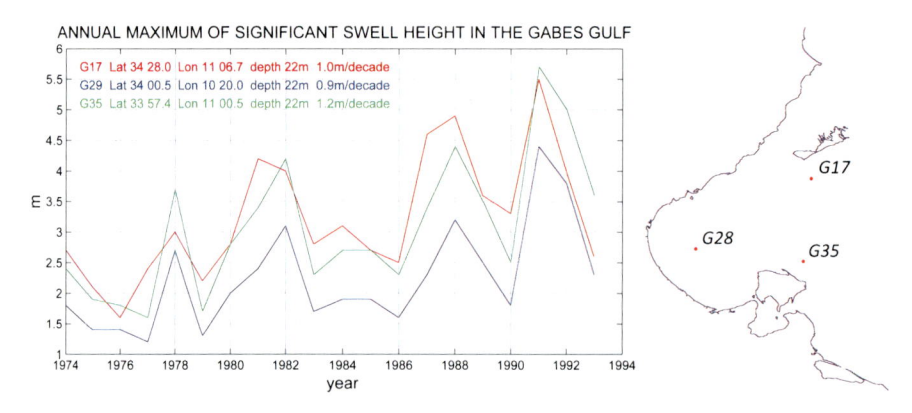

Fig. 4.14 Annual maximum significant swell height (m) 1974–1993 in three illustrative locations of the Gulf of Gabès (Data are from 'Direction Générale des Services Aériens et Maritimes', 'Ministère de l'Equipement, de l'Habitat et de l'Aménagement du Territoire', Tunisia, 1995)

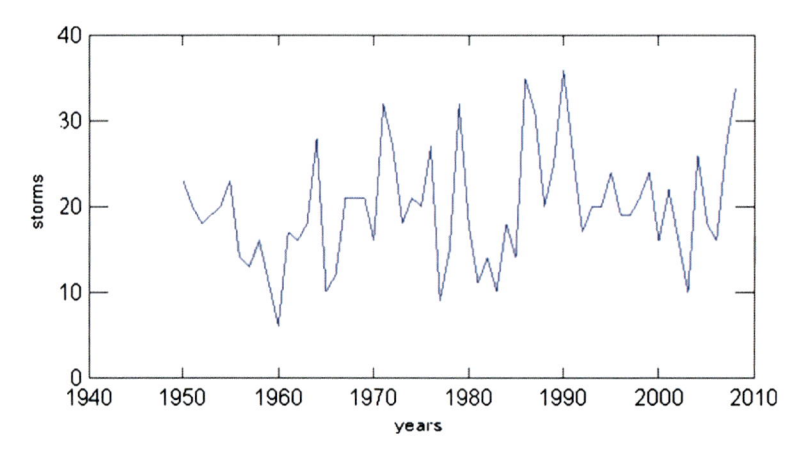

Fig. 4.15 Annual distribution of storms in Oran during the period 1950–2007. A storm event is defined according to the World Meteorological Organisation observation criteria of lightning and thunder storms reported at the same station (WMO 1992)

decade, which is almost double that for the longer period (1948–2008), 0.26°C/decade. Moreover the recent acceleration is stronger in summer – the trend for summers during the 1948–2008 period is 0.23°C/decade, which is close to the annual trend, whereas summers for the more recent period, 1973–2008, show a trend of 0.53°C/decade. However, such acceleration could be in part explained by the decadal variability, as show in the different plots.

As with all the CIRCE case studies, precipitation variability and trend were also considered a key hazard for the coastal case studies. However, records of precipitation from the coastal case-study locations show high inter- and intra-annual variability, typical of the Mediterranean region, with relatively small long-term trends.

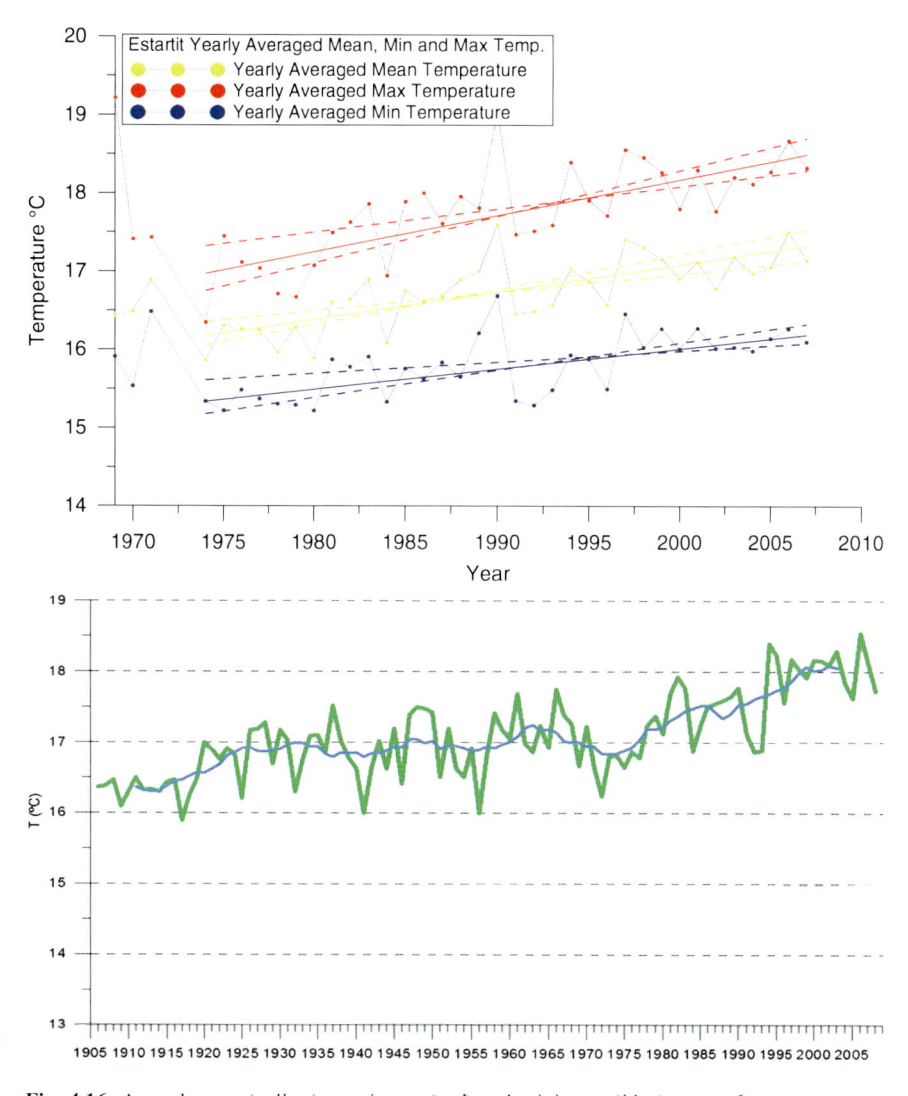

Fig. 4.16 Annual mean (*yellow*), maximum (*red*) and minimum (*blue*) sea surface temperature (SST) at the Northern Catalan Coast (Estartit), 1969–2008 (*top*) and, air temperature evolution, 1905–2008 (*bottom*)

The Chellif River is of major importance to the Gulf of Oran region. It is the longest river in Algeria, rising in the Saharian Atlas on the slopes of Djebel Amour and flowing for a distance of 800 km (a surface area of 43,750 km²) before discharging into the Mediterranean Sea at the Gulf of Arzew. In the Chellif River, monthly and annual flow regime is extremely variable, and in general, the hydrous river flow is noteworthy from September to May but practically negligible during the summer (Kies 2005). From the 1970s, the rainfall pattern in the basin of Chellif underwent

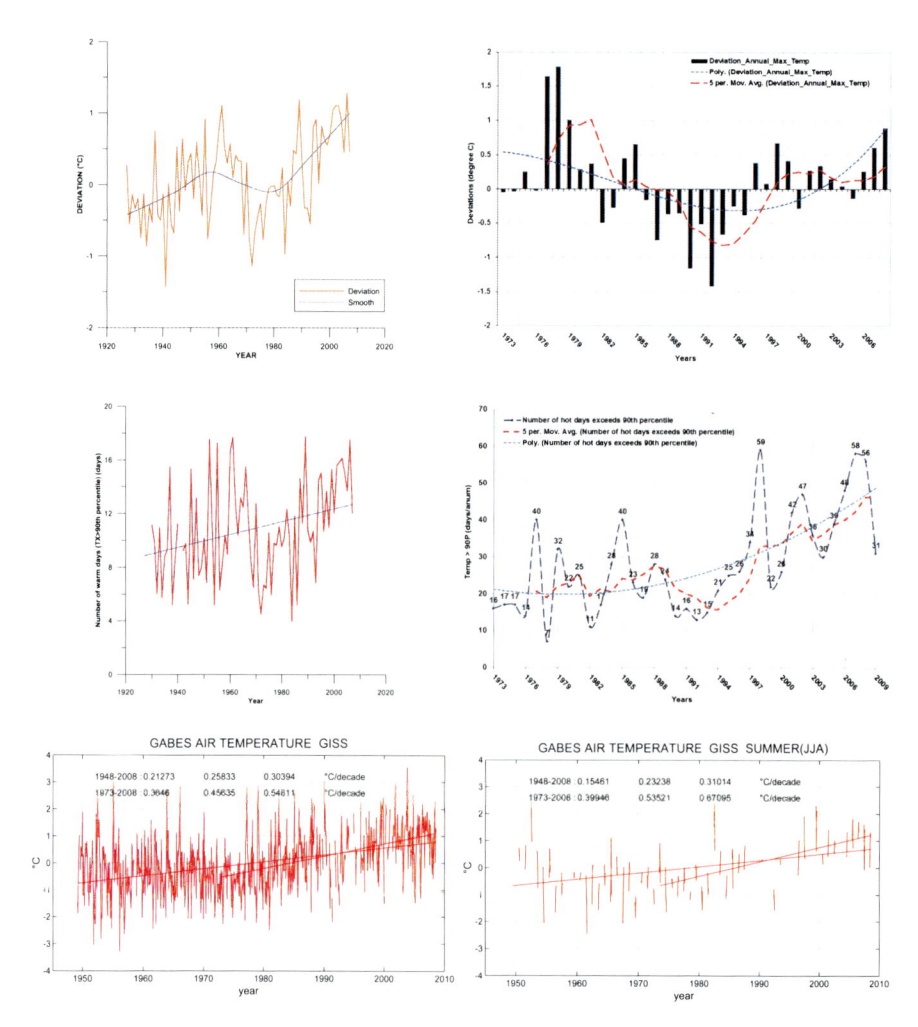

Fig. 4.17 Mean annual maximum temperature anomalies (from 1971 to 2000 average) for Gulf of Oran for 1927–2007 (*top left*) and West Nile Delta for 1973–2009 (*top right*). *Middle figures* show number of hot days, where maximum temperature exceeds the 1961–1990 90th percentile temperatures for Gulf of Oran (*left*) and West Nile Delta (*right*). *Bottom figures* show monthly anomalies of air temperature in the Gabès station for 1948–2008 based on data archived at the Goddard Institute for Space Studies (GISS, www.giss.nasa.gov). Trends with 95% confidence limits are shown for the period 1948–2008 and 1973–2008. The *bottom left figure* is annual and the *bottom right figure* is for the summer season only

a change that led to the emergence of a rainfall deficit, and this precipitation decline continued during the 1990s. Meddi and Boucefiane (2008) reported that the periods 1980–1994 and 1998–2001 are characterized by a significant deficit in rainfall recorded in various stations across the basin. Changes in outflow of the Chellif River can influence the salinity, dynamics and ecosystems within the Gulf of Oran region. Negative climate-related impacts on prey species distribution and abundance in this area, can have significant indirect consequences on food-web changes on odontocetes cetaceans such as common dolphin (*Delphinus delphis*) and bottlenose dolphin (*Tursiops truncates*) off the Algerian coast.

For tourist areas like Djerba Island in the southern Gulf of Gabès, the changing weather conditions may impact the attractiveness of the region. A Daily Climate-Tourism Potential Index (DCTPI) evaluates climate as a resource for tourism activity based on daily air temperature (mean, minimum and maximum), precipitation, relative humidity and wind speed. DCTPI (Henia and Alouane 2007) is a combined measure of the thermal comfort (in °C, Thom and Bosen 1959), the cooling power (Kcal/m^2, Siple and Passel 1945) and the weather conditions that influence outdoor activities. The thermal comfort (based on air temperature and humidity) and the cooling power (based on air temperature and wind speed) are expressed using indexes varying from 0 (less comfortable) to 3 (most comfortable). Three-hourly temperature values are usually used. However as only daily values are available for this location, daily maximum, minimum and mean values were used. Weather conditions that influence outdoor activities are expressed using a precipitation index (0 for daily precipitation >5 mm, 1 for daily precipitation from 2 to 5 mm and 2 for daily precipitation <2 mm) and a wind speed index (0 for wind speed >8 m/s, 1 for wind speed <8 m/s). Note that the precipitation index codes are slightly different from those defined in Henia and Alouane (2007), and sun radiation conditions are not used. The sum of the above indexes enables the definition of five classes of climate-tourism categories (DCTPI <2: highly unfavorable, 2–4: unfavorable, 4–6: acceptable, 6–8: favorable, >8: highly favorable). For more details on the calculation of DCTPI see: http://www.climato.be/aic/colloques/actes/carthage2007_actes.pdf.

During the 31-year period 1973–2003 (Fig. 4.18) some 'highly favorable' and 'favorable' days are replaced by 'acceptable' days (3.2% per decade). This change is most prominent in summer, and it reveals a potential reduction of the number of days with weather conditions favorable for tourism outdoor activities. However, it is noted that the lengthening of the summer 'tourist season' may largely balance this reduction. In addition, it must be remembered that climate is only one of many social, economic and environmental factors influencing tourism but which could not be quantified in this study.

4.4.3 Biogeophysical and Social Vulnerabilities

Biophysical vulnerability in Mediterranean coastal zones is largely influenced by:

(i) Lack of space for coastal systems to migrate inland due to high pressures of use and population density, and
(ii) Uniqueness of many threatened coastal ecosystems.

Whereas, social vulnerability is mainly related to:

(i) Low-lying urban areas with a high population density and buildings or public paths frequently exposed to direct wave action under the present climate;

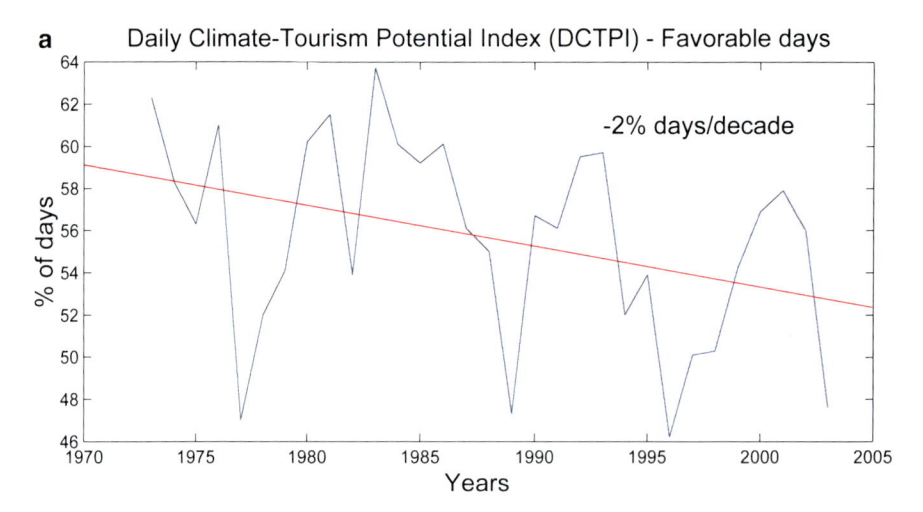

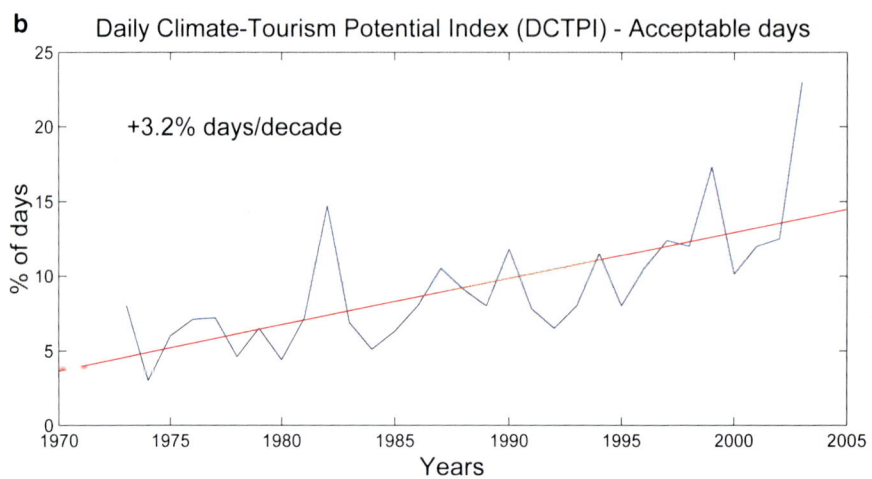

Fig. 4.18 Daily Climate Tourism Potential Index time series (1972–2003) for the classes 'favorable days' (**a**) and 'acceptable days' (**b**) for Djerba (Gulf of Gabès). *Linear trends* are shown with the trend rates (in % days/decade)

(ii) Infrastructure such as roads, railways or coastal engineering structures located in the coastal zone; and
(iii) Economic activities and facilities located in the coastal zone. They include: tourism, airports, water treatment plants, energy and industrial installations, and agriculture.

The coastal case studies have shown that both biogeophysical and social vulnerabilities in some Mediterranean coastal areas have significantly affected both coastal erosion processes as well as water availability (Table 4.6). The low-lying deltaic areas of the four CIRCE case study locations makes them among the most biogeophysically vulnerable locations around the Mediterranean to changes in sea level and/or wave

Table 4.6 Key biogeophysical and social vulnerabilities identified for the CIRCE coastal case study locations, Gulf of Gabès (Ga), Gulf of Oran (Or), Gulf of Valencia (Va) and West Nile Delta (WN)

Biogeophysical vulnerabilities	Social vulnerabilities
Coastal	**Coastal**
Area of land used as coastal buffer (Ga, Or, Va, WN)	Discharges/sources of pollution (Ga, Va, WN)
Coastal outflow of freshwater; nutrient load/fluxes (Ga, Or, Va, WN)	Population/Industry/Infrastructure/Tourism facilities location in coastal zone (Ga, Or, Va, WN)
	Coastal protection (e.g., expenditure, area protected; number of treaties) (Ga, Or, Va, WN)
Marine ecosystems and fisheries	**Marine ecosystems and fisheries**
Marine pollution/water quality (Ga, Or, Va, WN)	Density of population/industry/infrastructure/ located in the coastal zone (Ga, Or, Va, WN)
Fish stocks (Ga, Or, Va, WN)	Coastal protection expenditure/quotas/marine protection areas (Ga, Or, Va, WN)
Sardine, Saurel, Anchovy (Or)	Discharges/sources of pollution (Ga, Va, WN)
Coastal outflow of freshwater, nutrient load/fluxes (Ga, Or, Va, WN)	% employed in fishing industry/GDP (Ga)
Biodiversity, Alien species (Ga)	
Freshwater resources	**Freshwater resources**
Freshwater quality, nutrient load (Or, Va, WN)	Total consumption by sector (Ga, Or, Va, WN)
Salinization, salt wedges, freshwater aquifers vulnerable to inundation (Ga, Or, Va, WN)	Per capita use of freshwater (Ga, Or, Va, WN)
Runoff, river discharge (Ga, Or, Va, WN)	Water resources/environmental management (expenditure, area protected) (Ga, Or, Va, WN)
Abstraction (surface, groundwater) (Ga, Or, Va, WN)	Discharges, sources of pollution (WN)
Agriculture	**Agriculture**
Soil quality/erosion (Ga, Va, WN)	% employed in agriculture (Ga Va, WN)
Water availability (Ga, Va, WN)	% contribution of agriculture to GDP (Ga, Va, WN)
Salinization, inundation, shoreline retreat (Ga, Va, WN)	% agricultural land located on land vulnerable to flooding (Ga, WN)
Area of arable land (Ga, Va, WN)	
Transported air pollutants, dust (WN)	
Tourism	**Tourism**
Coastal environment (coastal erosion, shoreline retreat, inundation, pollution) (Ga, Or, Va)	% employed in tourism, contribution of tourism to GDP (Ga, Or, Va)
Change of favorable/unfavorable weather conditions in summer (excess warming)	GDP per capita (Ga, Or, Va)
	% tourism facilities located in coastal zone and vulnerable to flooding (Ga)

(continued)

Table 4.6 (continued)

Biogeophysical vulnerabilities	Social vulnerabilities
Health and wellbeing	**Health and wellbeing**
Vector habitat suitability (Ga, Or)	Per capita health expenditure, access to healthcare, patients per GP (Or, Va)
Air pollution (Or, Va)	% population <5 years >65 years (Ga, Or, Va)
	Health preparedness/emergency planning (Or)
Industry (energy, petrochemical)	**Industry** (energy, petrochemical)
Coastal erosion, retreat (Or)	% urban (industry, residential) area located in flood vulnerable zone (Or)
Inundation/coastal flooding (Or)	Coastal protection (e.g. expenditure, area protected, number of treatise) (Or)
Water availability (for industry) (Or)	% employed in industry, contribution of industry to GDP (Or)
Urban infrastructure (including energy)	**Urban infrastructure** (including energy)
Coastal erosion, shoreline retreat (Va)	% urban (industry, residential) area located in flood vulnerable zone (Va)
Inundation/coastal flooding (Va)	Coastal protection (e.g. expenditure, area protected, number of treatise (Va)

storminess, as well as to changes in aquifer regeneration and sedimentation processes. For example, increases in freshwater abstraction from the Ebro River (which feeds into the Gulf of Valencia-Catalan coast) for agricultural and domestic purposes, are associated with a reduction in river flow from an average annual discharge of 424 m^3/s (Sierra et al. 2002, 2004) between 1960 and 1990 (maximum monthly values in February of 662 m^3/s and minimum in August of 135 m^3/s) to an annual average of 300 m^3/s (461 and 119 m^3/s for the monthly average maximum and minimum values respectively), for the last decade of this period (1980–1990). Furthermore, the concomitant decrease in river sediment transportation to the Ebro delta is associated with increased rates of coastal erosion and subsidence, which further increase the vulnerability of the delta to sea-level rise and salinization.

A large number of tourist areas in the southern Mediterranean region are coastal and highly dependent on the quality of beaches, e.g., the Djerba tourist zone in the south of the Gulf of Gabès. These are characterized by low-lying lands (such as Kerkennah, Kneis and Djerba islands in the Gulf of Gabès) that are vulnerable to coastal erosion, sea-level rise and storms. Shallow marine areas (such as the Gulf of Gabès) also constitute vulnerable ecosystems – where their vulnerability is related to the rapid and strong response of the water body to atmospheric changes (e.g. the water rapidly warms as consequence of a warmer air temperature), and the associated biophysical impacts on the ecosystems. Since 1990, for example, the 'Grape algae' Chlorophyta *Caulerpa racemosa* var. *cylindracea* has colonized 12 Mediterranean countries (Libya, Greece, Italy, France, Turkey, Cyprus, Tunisia, Spain, Albania, Algeria, Croatia, Malta) (Klein and Verlaque 2008) and all the major Mediterranean islands (Balearic Islands, Corsica, Crete, Cyprus, Sardinia and Sicily) (Verlaque et al. 2003). While there are many other factors that may have contributed to this change, water temperature is one potential contributing factor.

Alien jellyfish species like the 'mauve stinger' (*Pelagia noctiluca*), which belong to the top 100 'worst invasive' species (Cheviron 2007), pose a threat to numerous native Mediterranean species in the Gulf of Oran. Invasion by alien marine species is also an issue in the Gulf of Gabès (see for example Fig. 3.7). However, other invasive species, such as Spanish mackerel (*Scomberomorus commerson*), have become economically important and have helped to reduce the economic vulnerability of the area. Regarding terrestrial coastal species, the nature reserve of La Macta in the Gulf of Oran is a site of special ecological interest, situated in a triangular depression separated from the Gulf of Arzew near Oran by a dune cordon, a complex of open water, marsh, and humid steppe some nine miles inland, supplied by three permanent rivers. The site represents a fragile wetland ecosystem that is rare in North Africa in terms of the diversity of habitats it contains, most notably the 'sansouires' salt marshes. Several vulnerability factors affect the ecological balance of the site and contribute to the destruction of its vegetation cover and the disappearance of its fauna, including erosion and the degradation of the eastern part of the dune system and biological invasion of the desert truffles (known in the region as *Terfess*).

The Mediterranean coastal zones have some of the highest population densities in the Mediterranean area, and support an unparalleled concentration of uses and infrastructure in the region. They are also vital for their high biological productivity and the link they provide between terrestrial and aquatic ecosystems. The examples discussed in this section have highlighted the complexity of these coastal systems, and the many environmental and social pressures that render them particularly vulnerable to weather/climate changes. In Chap. 5 these vulnerabilities are considered in the context of future projected climate changes and their associated natural hazards, and the options to adapt coastal systems and activities to current and future projected climate changes are considered in Chap. 6.

References

Afik T (2009) Quantitative estimation of CO_2 fluxes in a semi-arid forest and their dependence on climatic factors. Thesis, The Hebrew University of Jerusalem, Rehovot

Agnoletti M (2007) The degradation of traditional landscape in a mountain area of Tuscany during the 19th and 20th centuries: implications for biodiversity and sustainable management. For Ecol Manag 249:5–17

Agnoletti M, Paoletti S, Marinai V (2006) Paesaggio Regione Toscana, Segnali Ambientali 2006. EDIFIR, Firenze, pp 47–51

Balestrieri G (ed) (2005) Il turismo rurale nello sviluppo territoriale integrato della Toscana. IRPET-Regione Toscana, Florence

Bernetti I, Marinelli N (2010) Evaluation of landscape impacts and land use change: a Tuscan case study for CAP reform scenarios. Aestimum 56, Giugno 2010:1–29

Bernetti I, Franciosi C, Lombardi G (2006) Land use change and the multifunctional role of agriculture: a spatial prediction model in an Italian rural area. Int J Agric Res 5(2/3):145–161

Checkley W, Epstein L, Gilman R, Figueroa D, Cama R, Patz J, Black R (2000) Effects of El Nino and ambient temperature on hospital admissions for diarrheal diseases in Peruvian children. Lancet 355:442–450

Cheviron N (2007) Le réchauffement climatique menace la Méditerranée. www.cyberpress.ca

Chou W-C, Wu J-L, Wang Y-C, Huang H, Sung F-C, Chuang C-Y (2010) Modeling the impact of climate variability on diarrhea-associated diseases in Taiwan (1996–2007). Sci Total Environ 409:43–51

Court-Picon M, Gadbin-Henry C, Guibal F, Roux M (2004) Dendrometry and morphometry of *Pinus pinea* L. in Lower Provence (France): adaptability and variability of provenances. For Ecol Manag 194:319–333

Dawson IA, Wardlaw IF (1989) The tolerance of wheat to high temperatures during reproductive growth. III. Booting to anthesis. Aust J Agr Res 40(5):965–980

De Wrachien D, Ragab R, Giordano A (2003) Climate change land degradation and desertification in the Mediterranean environment. In: Proceedings of the NATO – CCMs and science committee workshop on desertification in the Mediterranean region. NATO security through science series, 3/2006, pp 353–371, doi:10.1007/1-4020-3760-0_16

EEA (2003) Europe's water: an indicator-based assessment. Topic Report, European Environment Agency, Copenhagen, 97 pp

El-Zein A, Tewtel-Salem M (2005) On the association between high temperature and mortality in warm climates. Sci Total Environ 343:273–275

El-Zein A, Tewtel-Salem M, Nehme G (2004) A time-series analysis of mortality and air temperature in Greater Beirut. Sci Total Environ 330:71–80

Evans JP (2009) 21st century climate change in the Middle East. Clim Chang 92:417–432

Flannigan M, Campbell I, Wotton M, Carcaillet C, Richard P, Bergeron Y (2001) Future fire in Canada's boreal forest: paleoecology results and general circulation model—regional climate model simulations. Can J For Res 31:854–864

Fleischer A, Sternberg M (2006) The economic impact of global climate change on Mediterranean rangeland ecosystems: a space-for-time approach. Ecol Econ 59:287–295

Founda D, Giannakopoulos C (2009) The exceptionally hot summer of 2007 in Athens, Greece. A typical summer in the future climate. Glob Planet Chang 67:227–236

Giannakopoulos C, LeSager P, Bindi M, Moriondo M, Kostopoulou E, Goodess C (2009) Climatic changes and associated impacts in the Mediterranean resulting from a 2°C global warming. Glob Planet Chang 68(3):209–224

Giorgi F (2006) Climate change hot-spots. Geophys Res Lett 33:L08707

Giorgi F, Lionello P (2008) Climate change projections for the Mediterranean region. Glob Planet Chang 63:90–104

Grünzweig JM, Gelfand I, Fried Y, Yakir D (2007) Biogeochemical factors contributing to enhanced carbon storage following afforestation of a semi-arid shrubland. Biogeosciences 4:891–904

Grünzweig JM, Hemming D, Maseyk K, Lin T, Rotenberg E, Raz-Yaseef N, Falloon PD, Yakir D (2009) Water limitation to soil CO_2 efflux in a pine forest at the semi-arid 'timberline'. J Geophys Res 114:G03008

Hashizume M, Armstrong B, Hajat S, Wagatsuma Y, Faruque ASG, Hayashi T, Sack DA (2007) Association between climate variability and hospital visits for non-cholera diarrhea in Bangladesh: effects and vulnerable groups. Int J Epidemiol 35(5):1030–1037

Henia L, Alouane T (2007) Le potentiel climato-touristique de la Tunisie, Actes du X Xème Colloque de l'Association de Climatologie, Carthage, 3–8 Septembre 2007

Holzapfel C, Tielbörger K, Parag HA, Kigel J, Sternberg M (2006) Annual plant–shrub interactions along an aridity gradient. Basic Appl Ecol 7:268–279. doi:10.1016/j.baae.2005.08.003

Howden SM, Moore JL, McKeon GM, Reyenga PJ, Carter JO, Scanlan JC (1999) Global change impacts on fire dynamics in the mulga woodlands of south-west Queensland. Working paper series 99/05. CSIRO Wildlife and Ecology, Canberra. Available at www.cse.csiro.au/publications/1999/firemulga-99-05.pdf

Iliadis LS (2004) A decision support system applying an integrated fuzzy model for long-term forest fire risk estimation. Environ Model Software 20:613–621

IPCC (2007) In: Parry ML, Canziani OF, Palutikof JP, van der Linden PJ, Hanson CE (eds) Climate change 2007: impacts, adaptation and vulnerability. Contribution of working group II to the fourth assessment report of the intergovernmental panel on climate change. Cambridge University Press, Cambridge, 976 pp

IRPET (2009) Toscana CO_2. Prime valutazioni sulla sfida dei cambiamenti climatici. Florence

JNF (2002) Report on drought damage in forests of the southern district. Jewish National Fund, Gilat, Israel (in Hebrew)

Khair K, Aker N, Zahruddine K (1992) Hydrogeologic units of Lebanon. Hydrogeol J 1(2):34–49

Kies F (2005) Influence de l'oued Cheliff sur l'écosystème marin dans la zone de l'embouchure (wilaya de Mostaganem). Ingéniorat d'Etat en Ressources Halieutiques. Université Abdelhamid Ibn Badis de Mostaganem, Algérie

Klein J, Verlaque M (2008) The *Caulerpa racemosa* invasion: a critical review. Mar Pollut Bull 56:205–225

Kurz WA, Apps MJ, Stocks BJ, Volney JA (1995) Global climate change: disturbance regimes and biospheric feedbacks of temperate and boreal forests. In: Woodwell GM, Mackenzie FT (eds) Biotic feedbacks in the global climatic system. Will the warming feed the warming? University Press, New York, pp 119–133

Lavorel S, Canadell J, Rambal S, Terradast J (1998) Mediterranean terrestrial ecosystems: research priorities on global change effects. Glob Ecol Biogeogr Lett 7(1):57–166

Lionello P et al (2006) The Mediterranean climate: an overview of the main characteristics and issues. In: Lionello P, Malanotte-Rizzoli P, Boscolo R (eds) Mediterranean climate variability. Elsevier, Amsterdam, pp 1–26

Long SP (1991) Modification of the response of photosynthetic productivity to rising temperature by atmospheric CO_2 concentrations: has its importance been underestimated? Plant Cell Environ 14(8):729–739

Luterbacher J et al (2006) Mediterranean climate variability over the last centuries. A review. In: Lionello P, Malanotte-Rizzoli P, Boscolo R (eds) Mediterranean climate variability. Elsevier, Amsterdam, pp 27–148

McCarthy MP (2009) Report on the CIRCE urban heat island simulations, deliverable D11.3.3 of the sixth framework program, Project No. 036961-CIRCE. Available from http://www.cru.uea.ac.uk/projects/circe/urban.html

McCarthy MP, Best MJ, Betts RA (2010) Climate change in cities due to global warming and urban effects. Geophys Res Lett 37:L09705. doi:10.1029/2010GL042845

Mearns LO, Rosenzweig C, Goldberg R (1996) The effect of changes in daily and interannual climatic variability on CERES-Wheat: a sensitivity study. Clim Chang 32(3):257–292

Meddi M, Boucefiane A (2008) Impact of climate change on the renewal of groundwater resources in the Chéliff-Zahrez basin of Algeria. International conference groundwater & climate in Africa, Kampala, Uganda, 24–28 June 2008

Morabito M, Crisci A, Barcaioli G, Maracchi G (2004) Climate change: the impact on tourism comfort at three Italian tourist sites. In: Matzarakis A, de Freitas C, Scott R (eds) Advances in tourism climatology. Ber. Meteor. Inst. Univ. Freiburg, pp 56–65

Morabito M, Cecchi L, Modesti PA, Crisci A, Orlandini S, Maoracchi GD, Gensini GF (2005) The impact of hot weather conditions on tourism in Florence, Italy: the summer 2002–2003 experience. In: Matzarakis A, de Freitas CR, Scott D (eds) Advances in tourism climatology. Meteorological Institute of the University of Freiburg, Freiberg

Mosso C, Mestres M, Sierra JP, Sánchez-Arcilla A, Goodess C (2009) Waves and surges in the Valencia gulf. Variability rather than climate change. J Coast Res 56(SI):248–251

Natali F, Dalla Marta A, Orlando F, Orlandini S (2009) Water use in Italian Agriculture: analysis of rainfall pattern irrigation system and water storage capacity of farm ponds. Ital J Agron 55(3)

Nuvoli S, Matina P (2002) Agricoltura sostenibile e consumi idrici. VII Conferenza Regionale sull'Ambiente 2002: Strategie per la sostenibilità e integrazione delle politiche–Sessione Ambiente e Agricoltura

Oren A, Steinberger Y (2008) Coping with artifacts induced by $CaCO_3$–CO_2–H_2O equilibria in substrate utilization profiling of calcareous soils. Soil Biol Biochem 40:2569–2577

Price C, Stone L, Huppert A, Rajagopalan B, Alpert P (1998) A possible link between El Nino and precipitation in Israel. Geophys Res Lett 25:3963–3966

Rotenberg E, Yakir D (2010) Contribution of semi-arid forests to the climate system. Science 327:451–454

Saadeh M (2008) Influence of overexploitation and seawater intrusion on the quality of groundwater in greater Beirut. PhD thesis, RWTH Aachen University, Germany

Sánchez-Arcilla A, Mösso C, Sierra JP, Mestres M, Harzallah A, Senouci M, El Raey M, El-Askary H (2010) Climatic drivers of potential hazards in Mediterranean coasts. Reg Environ Chang 11:617–636. doi:10.1007/s10113-010-0193-6

Sierra JP, Sánchez-Arcilla A, González Del Río J, Flos J, Movellán E, Mösso C, Martínez R, Rodilla M, Falco S, Romero I (2002) Spatial distribution of nutrients in the Ebro estuary and plume. Cont Shelf Res 22(2):361–378, ISSN: 0278-4343

Sierra JP, Sánchez-Arcilla A, Figueras PA, González Del Río J, Rassmusen EK, Mösso C (2004) Effects of discharge reductions on salt wedge dynamics of the Ebro river. River Res Appl 20(1):61–77, ISSN: 1535-1459

Singh R, Hales S, Wet N, Raj R, Hearnden M, Weinstein P (2001) The influence of climate variation and change on diarrheal disease in the Pacific Islands. Environ Health Perspect 109(2):155–159

Siple PA, Passel CF (1945) Measurements of dry atmospheric cooling in subfreezing temperatures. In: Proceedings of the American Philosophical Society, vol 89(1), Reports on scientific results of the United States Antarctic Service Expedition, 1939–1941 (30 April 1945), Published by American Philosophical Society, Philadelphia, pp 177–199

SOGREAH Engineering (1998) Etude de l'assainissement du groupement urbain d'Oran, Mission A: actualisation du plan directeur d'assainissement PDAA. AGEP, Algérie

Talmon Y, Sternberg M, Grünzweig JM (2011) Impact of rainfall manipulations and biotic controls on soil respiration in Mediterranean and desert ecosystems along an aridity gradient. Glob Chang Biol 17(2):1108–1118. doi:10.1111/j.1365-2486.2010.02285.x

Tashiro T, Wardlaw IF (1990) The effect of high temperature at different stages of ripening on grain set, grain weight and grain dimensions in the semi-dwarf wheat 'Banks'. Ann Bot 65:51–61

Thom EC, Bosen JF (1959) The discomfort index. Weatherwise 12:57–60

Tubiello FN, Donatelli M, Rosenzweig C, Stockle CO (2000) Effects of climate change and elevated CO_2 on cropping systems: model predictions at two Italian locations. Eur J Agron 13:179–189

Ulbrich U et al (2006) The Mediterranean climate change under global warming. In: Lionello P, Malanotte-Rizzoli P, Boscolo R (eds) Mediterranean climate variability. Elsevier, Amsterdam, pp 398–415

Verlaque M, Durand C, Huisman JM, Boudouresque CF, Le Parco Y (2003) On the identity and origin of the Mediterranean invasive *Caulerpa racemosa* (Caulerpales, Chlorophyta). Eur J Phycol 38:325–339

Weinstein A (1989) Phenology of *Pinus halepensis* at two sites in Israel. For Ecol Manag 26(4):305–309. doi:10.1016/0378-1127(89)90091-1

World Meteorological Organization (1992) International meteorological vocabulary, WMO no. 182. WMO, Geneva

Yakir D, Lev-Yadun S, Zangvil A (1996) El Niño and tree growth near Jerusalem over the last 20 years. Glob Chang Biol 2(2):97–101. doi:10.1111/j.1365-2486.1996.tb00054.x

Chapter 5
Integration of the Climate Impact Assessments with Future Projections

Clare M. Goodess, Maureen D. Agnew, Christos Giannakopoulos, Debbie Hemming, Skander Ben Salem, Marco Bindi, Mohamed Nejmeddine Bradai, Letizia Congedi, Camilla Dibari, Hesham El-Askary, Mutasem El-Fadel, Mohamed El-Raey, Roberto Ferrise, Dimitra Founda, José M. Grünzweig, Ali Harzallah, Maria Hatzaki, Gillian Kay, Piero Lionello, César Mösso Aranda, Theib Oweis, Joan Pau Sierra, Basil Psiloglou, Marco Reale, Agustín Sánchez-Arcilla, Mohamed Senouci, Annalisa Tanzarella, and Konstantinos V. Varotsos

Abstract Climate projections are essential in order to extend the case-study impacts and vulnerability assessments to encompass future climate change. Thus climate-model based indicators for the future (to 2050 and for the A1B emissions scenario) are presented for the climate and atmosphere theme (including indices of temperature and precipitation extreme events), together with biogeophysical and socioeconomic indicators encompassing the other case-study themes. For the latter, the specific examples presented here include peri-urban fires, air pollution, human health risks, energy demand, alien marine species and tourism (attractiveness and socio-economic consequences). The primary source of information about future climate is the set of global and regional model simulations performed as part of CIRCE. These have the main novel characteristic of incorporating a realistic representation of the Mediterranean Sea including coupling between sea and atmosphere. These projections are inevitably subject to uncertainties relating to unpredictability, model structural

---

C.M. Goodess (✉) • M.D. Agnew
Climatic Research Unit, School of Environmental Sciences,
University of East Anglia, UEA, Norwich, UK
e-mail: c.goodess@uea.ac.uk

C. Giannakopoulos • D. Founda • M. Hatzaki • B. Psiloglou • K.V. Varotsos
Institute for Environmental Research and Sustainable Development,
National Observatory of Athens, Athens, Greece

D. Hemming • G. Kay
Climate Impacts Analysis Team, Met Office Hadley Centre, Exeter, Devon, UK

M. Bindi • R. Ferrise
Department of Plant, Soil and Environmental Science (DIPSA),
University of Florence, Florence, Italy

Institute for Biometeorology, National Research Council of Italy,
(CNR-IBIMET) Florence, Italy

S. Ben Salem • M.N. Bradai • A. Harzallah
Institut National des Sciences et Technologies de la Mer – INSTM Salammbô, Tunisia

uncertainty and value uncertainty. These uncertainties are addressed by taking a multi-model approach, but problems remain, for example, due to a systematic cold bias in the CIRCE models. In the context of the case-study integrated assessments, there are also uncertainties 'downstream' of climate modeling and the construction of climate change projections – largely relating to the modeling of impacts. In addition, there are uncertainties associated with all socio-economic projections used in the case studies – such as population projections. Thus there are uncertainties inherent to all stages of the integrated assessments and it is important to consider all these aspects in the context of adaptation decision making.

Keywords Climate change • Mediterranean • Climate projections • Impacts • Integrated assessment

---

L. Congedi
Centro Euro Mediterraneo per i Cambiamenti Climatici (CMCC), Lecce, Italy

C. Dibari
Department of Plant, Soil and Environmental Science (DIPSA),
University of Florence, Florence, Italy

H. El-Askary
School of Earth and Environmental Sciences, Schmid College of Science and Technology, Chapman University, Orange, CA, USA

Department of Environmental Sciences, Faculty of Science,
Alexandria University Alexandria, Egypt

M. El-Fadel
Faculty of Engineering and Architecture, American University of Beirut, Beirut, Lebanon

M. El-Raey
University of Alexandria, Alexandria, Egypt

J.M. Grünzweig
Institute of Plant Sciences and Genetics in Agriculture, The Robert H. Smith
Faculty of Agriculture, Food and Environment, The Hebrew University of Jerusalem,
Jerusalem, Israel

P. Lionello
Centro Euro Mediterraneo per i Cambiamenti Climatici (CMCC), Lecce, Italy

Department of Material Science, University of Salento, Lecce, Italy

C. Mösso Aranda • J. Pau Sierra • A. Sánchez-Arcilla
Laboratori d'Enginyeria Marìtima (LIM/UPC), Universitat Politécnica de Catalunya,
Barcelona, Spain

T. Oweis
International Center for Agricultural Research in the Dry Areas, ICARDA, Aleppo, Syria

M. Reale • A. Tanzarella
Department of Material Science, University of Salento, Lecce, Italy

M. Senouci
Association Recherche Climat et Environnement – ARCE, Oran, Algeria

5.1 Introduction

In the previous chapter, impacts and vulnerability due to hazards associated with current climate variability and change are discussed. Understanding the current vulnerability provides the context for considering future changes which are the focus of this chapter. In accordance with the CCIF described in Chap. 1, the CIRCE case studies aimed to construct a set of indicators describing the response to changes in emissions of greenhouse gases, climate variability and change. Chapter 3 describes the indicator selection process encompassing both physical and socio-economic systems, categorized according to seven key themes (see Table 3.1). Here, climate-model based indicators for the future (to 2050) are presented in Sect. 5.2 for the Climate and atmosphere theme, while biogeophysical and socioeconomic indicators encompassing the other themes are discussed in Sect. 5.3. In each case, relevant work undertaken across the CIRCE project is summarized, before describing the case-study focused work. Section 5.4 discusses the uncertainties associated with these projections, together with the other sources of uncertainty arising at each step of this end-to-end assessment.

5.2 Climate Projections

Climate projections are essential in order to extend the impacts and vulnerability assessments presented in Chap. 4 to encompass future climate change. Global Climate Model (GCM) simulations provide the basis of regional projections, together with dynamically [using Regional Climate Models (RCMs)] or statistically downscaled information (Carter et al. 2007; Christensen et al. 2007). As part of the CIRCE project, a set of GCM and RCM simulations has been performed with the main novel characteristic of incorporating a realistic representation of the Mediterranean Sea including coupling between the sea and atmosphere (see Part I, Chap. 3). These simulations (summarized here in Table 5.1) provide the principle source of information about future climate for the CIRCE case studies, although supporting RCM-based evidence from the earlier EU ENSEMBLES project is also used, together with the published literature.

The six CIRCE models are all coupled atmosphere–ocean models. The justification of this experimental set up is described in Part I, Chap. 3, together with details of the modeling components. The spatial resolution of the Mediterranean Sea ocean components is 7–12 km (see Table 3.1 of Part I, Chap. 3). The atmospheric component of the global INGV model has a spatial resolution of about 80 km (Table 5.1), while CNRM is a stretched-grid global model, zooming in to a resolution of 50 km over Europe, the Mediterranean and North Africa. The other models are regional models with a spatial resolution of 25–30 km, with boundary conditions taken from coarser global models. The MPI regional model is forced by the CIRCE INGV model while the ENEA regional model is forced by a non-CIRCE run of the

Table 5.1 Summary of the CIRCE models

Acronym used in this chapter	Other acronyms used	Type of model, spatial resolution of the atmospheric component and boundary forcing	Seasonal mean response over the Mediterranean for 2021–2050 with respect to 1961–1990			
			Tdjf	Tjja	Pdjf	Pjja
CNRM		Stretched grid global model zooming to 50 km over the Mediterranean	0.78	1.42	−3	−3
INGV	CMCC-Med	Global model: ~80 km	1.44	1.68	−23	−8
IPSLreg	IPSL1 LMDreg	LMDZ regional model: 30 km. Forcing from an earlier (non-CIRCE) IPSL GCM run	2.01	2.46	−10	−5
IPSLglo	IPSL2 LMDglo	Coupled LMDZ global and LMDZ regional model: 30 km	1.00	1.16	+6	−6
ENEA	PROTHEUS	Regional model: 30 km Forcing from an ECHAM5 GCM (non-CIRCE) run	0.97	1.41	−15	+6
MPI		Regional model: 25 km. Forcing from INGV	1.38	1.52	−3	+1

See Part I, Chap. 3, Tables 3.1 and 3.14 for more details
Temperature (T) changes (°C) and Precipitation (P) changes (mm per season) for winter (djf) and summer (jja) are averages for all model land and sea grid boxes over the Mediterranean

ECHAM5 global model. Two configurations were used by IPSL: for IPSLreg (also referred to as LMDreg), the LMDZ regional model was driven by a non-CIRCE run of the IPSL global model, whereas for IPSLglo (also referred to as LMDglo) the LMDZ global and regional models were fully coupled. The simulations were all run for 1951–2050 (with the INGV simulation extended to 2100) using the IPCC-SRES A1B emissions scenario. Across the CIRCE project, results are generally presented as the difference between the 30-year baseline current period (1961–1990) and the future scenario period (2021–2050).

The CIRCE models are consistent in indicating warming over the Mediterranean, ranging from 0.8 to 2.0°C in winter (DJF) and from 1.2 to 2.5°C in summer (JJA) for 2021–2050 compared with 1961–1990 (Table 5.1). As well as the seasonal variability in warming (the strongest warming is in summer and weakest in winter, with intermediate values in spring and autumn), the CIRCE models also indicate spatial variability in warming (Part I, Sect. 3.2.1). In winter, the multi-model mean has a large maximum extending from the eastern part of the Balkans into central Russia with secondary maxima over western North Africa and the Middle East. In summer, maximum warming is projected over the Iberian Peninsula and the western part of North Africa.

The projections for precipitation are less consistent, reflecting the generally higher uncertainties for this variable (see Sect. 5.4). For winter, five of the CIRCE models indicate a decrease in precipitation over the whole region, varying from −3 to −23 mm per season while the IPSLglo model gives an increase (Table 5.1). For summer, four models project a decrease (ranging from −3 to −8 mm per season) and two indicate an increase. The projected increases in winter precipitation are, however, generally restricted to parts of the Northern Mediterranean and the Alps – and while uncertainties are larger for summer, the general picture is one of reduced summer precipitation (though it should be noted that summer rainfall is very low over much of the region).

Thus, in general, the six CIRCE models show a tendency towards warmer and drier conditions over the Mediterranean, consistent with earlier studies (e.g., Giorgi and Lionello 2008). Changes in temperature and precipitation extremes projected by the CIRCE models have also been explored (see Part I, Sect. 3.4). For temperature extremes, an analysis of five of the models (CNRM, INGV, IPSLreg, ENEA, MPI – see Part I, Sect. 3.4.3) indicates:

- a general increase in the number of very hot days and nights, together with longer warm spells and heat waves,
- largest increases over the Iberian Peninsula in summer
- general decreases in the number of very cold days and nights and shorter cold spells.

For precipitation extremes, a Mediterranean-wide analysis of changes in extended winter (October to March) heavy precipitation (based on 5-year return values) in the INGV global model (see Part I, Sect. 3.4.3) indicates:

- an increase of 5-year return values over the Iberian Peninsula (except the south), southern Italy and the Aegean (Part I, Fig. 3.15)

- a weak general increase in the percentage of precipitation associated with heavy precipitation events
- a decrease in the number of heavy precipitation days over the western and central Mediterranean, and an increase over the northeast (Part I, Fig. 3.16)

An unpublished analysis of extreme precipitation (defined using various indices including the 95th percentile of precipitation amounts) in five of the CIRCE models shows considerable inter-model variation in response. However, a few coherent changes are indicated, including a coherent increase in the winter 95th percentile for the central Iberian Peninsula.

Finally, analysis of Mediterranean winter cyclones and wind storms simulated by the INGV and MPI models (the latter regional model is driven by the former global model) and of wind simulated by an ensemble of seven GCMs including INGV and ECHAM5 (see Part I, Sect. 3.4.3) indicates:

- a general decrease in the density of cyclone tracks
- the only exception to the above decrease is a small area of increase between southern Turkey and Cyprus – one of the most important regions for cyclogenesis in the eastern Mediterranean
- a general decrease in the frequency of cyclones associated with extreme winds
- the only exception to the above decrease is some countries bordering the northern Mediterranean coast from Croatia to Greece which show a small increase in the number of wind storms
- the projected changes in winter cyclone activity and sea level pressure are consistent with the projected decrease in winter precipitation
- no significant change in the intensity of the most extreme wind storms at least until the end of the current century.

As part of the case-study work, climate indicators were constructed using output from the six CIRCE climate models. For the urban case studies, the nearest land-based model grid box was used (after checking that the simulated values were not inconsistent or anomalous compared with neighboring grid boxes). For the rural and coastal case studies, the average of land-based model grid boxes covering each region was used (again checking visually for any spatial inconsistencies). Daily data were used on the original grids (different for each model), while monthly temperature and precipitation were interpolated to a regular half-degree latitude-longitude grid. Where possible, simulated data were compared with available observed station data, and/or the gridded E-OBS data (Haylock et al. 2008). Such comparisons reveal a systematic cold bias in the CIRCE models of about 2°C across the Mediterranean. In some cases, bias correction was applied before using the data. See Sect. 5.4 for further discussion of model biases and correction techniques.

Mean annual maximum and minimum temperature (Tmax and Tmin) were used as core climate indicators for many of the case studies (see Sect. 3.3.1). The projected changes (for 2021–2050 with respect to 1961–1990) in these indicators are presented in Figs. 5.1 and 5.2 for five of the models (all except IPSLglo). The ensemble-mean change is shown, together with the changes for the five individual models. For both

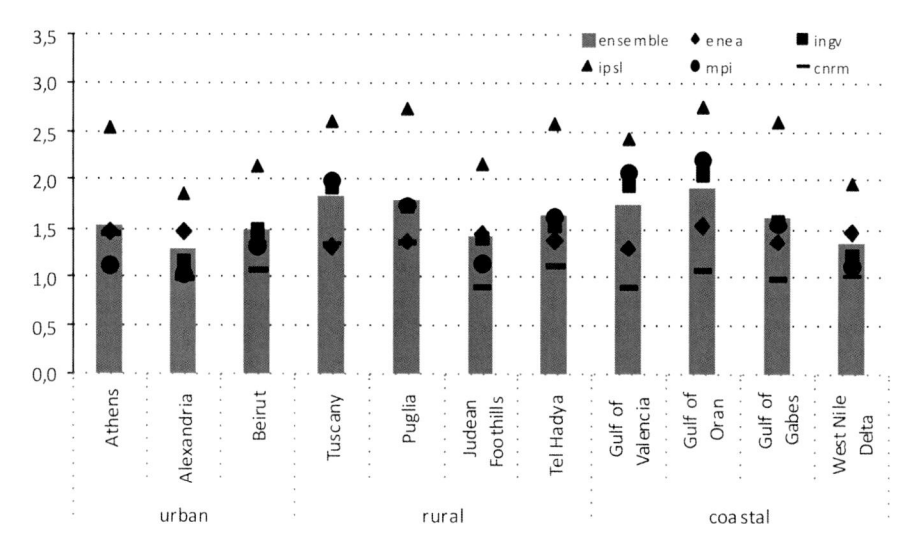

Fig. 5.1 Projected changes in annual Tmax (°C) (2021–2050 minus 1961–1990) based on CIRCE model data for the urban, rural and coastal case studies. The ensemble-mean change is shown, together with changes for the five individual models. IPSL is IPSLreg

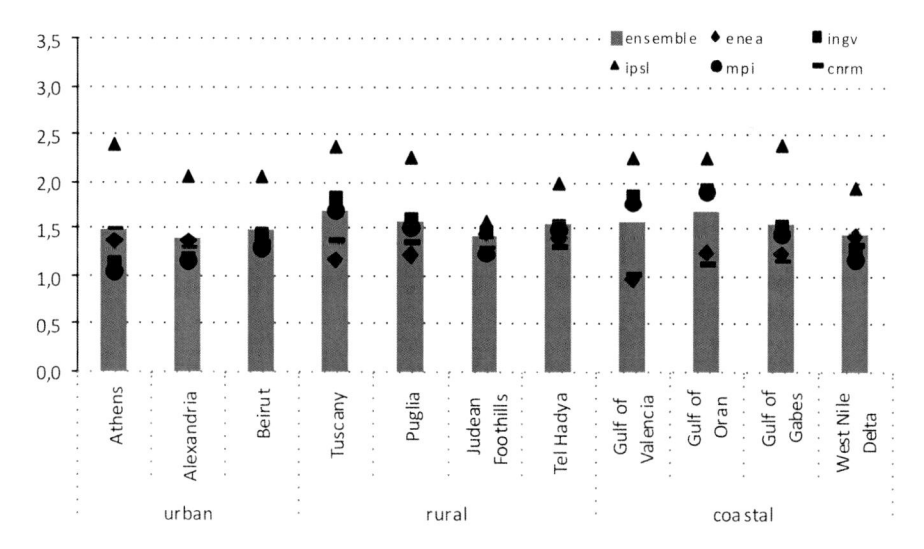

Fig. 5.2 Projected changes in Tmin (°C) (2021–2050 minus 1961–1990) based on CIRCE model data for the urban, rural and coastal case studies. The ensemble-mean change is shown, together with changes for the five individual models. IPSL is IPSLreg

Tmax and Tmin, the projected changes are slightly larger for case studies located in the western (Gulf of Valencia and Gulf of Oran) and central (Tuscany and Apulia) Mediterranean than in the eastern Mediterranean. Changes in Tmax are larger than changes in Tmin, particularly in the western Mediterranean. The largest ensemble-mean

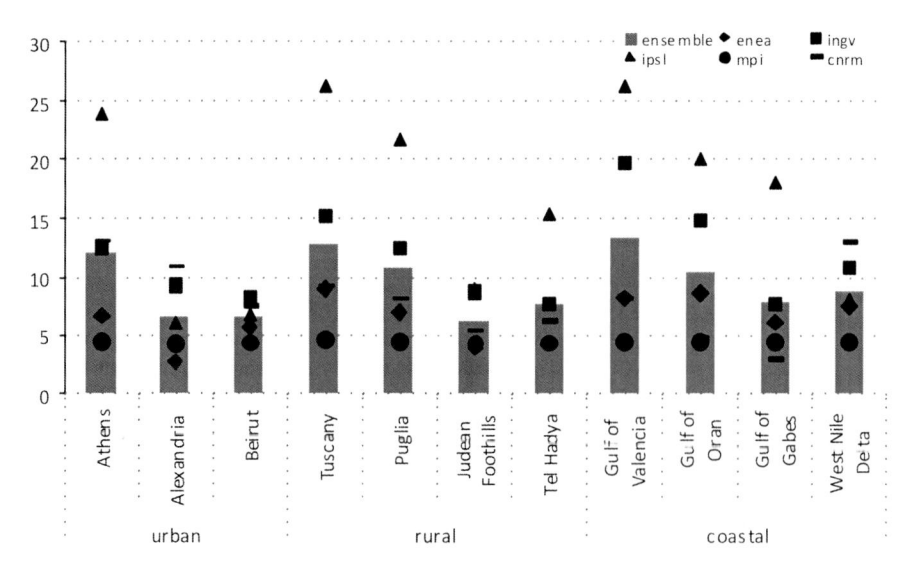

Fig. 5.3 Projected changes (2021–2050 minus 1961–1990) in the number of very hot summer days (Tx95n) based on CIRCE model data for the urban, rural and coastal case studies. The ensemble-mean change is shown, together with changes for the five individual models. IPSL is IPSLreg

change is for Tmax in the Gulf of Oran (almost 2°C). Figures 5.1 and 5.2 reflect the ensemble spread which is also identified in the Mediterranean-wide CIRCE analysis (see Part I, Chap. 3). The IPSLreg model consistently gives the greatest warming (more than 2.5°C for Tmax for many of the case studies), while the CNRM and ENEA models tend to give least warming (a little less than 1°C in some cases). The changes indicated by the MPI regional model and the INGV global model which provides its forcing are generally similar, though in a number of cases (particularly relating to extremes, see below) the projected regional model changes are smaller.

A number of indices of extremes were also selected as case-study specific climate indicators (Sect. 3.3.2). The Mediterranean-wide analysis of projected changes in temperature extremes (see above and Part I, Sect. 3.4) was therefore extended to focus on the case studies. Projected changes in very hot summer days (Tx95n: the number of days exceeding the 95th percentile of Tmax calculated for 1961–1990) are shown in Fig. 5.3 for all case studies. Reflecting the spatial patterns identified above (see also Part I, Fig. 3.13), the largest ensemble-mean change is for the Gulf of Valencia (about 13 more very hot summer days each year) and the smallest changes in the southeast Mediterranean (about six more very hot summer days for the Judean Foothills). The inter-model spread is very large – with the IPSLreg model indicating more than 25 additional very hot summer days for Tuscany and the Gulf of Valencia compared with less than five for the MPI model.

The number of very hot nights (Tn95n: the number of days exceeding the 95th percentile of Tmin calculated for 1961–1990) was selected as an indicator for the urban case studies since high night-time temperatures impact on human comfort. Projected changes for each season are shown in Fig. 5.4 (note that the threshold is

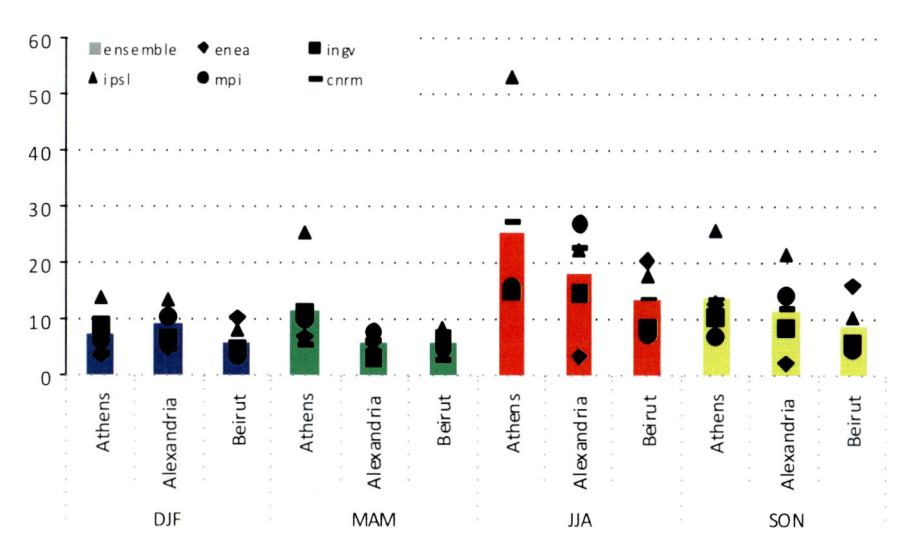

Fig. 5.4 Projected seasonal changes (2021–2050 minus 1961–1990) in the number of very hot nights (Tn95n) based on CIRCE model data for the urban case studies. The ensemble-mean change is shown, together with changes for the five individual models. IPSL is IPSLreg

calculated separately for each season). The largest changes are projected in summer and for Athens. The ensemble-mean change for Athens in summer is about 25 additional very hot nights, though this value is influenced by the anomalously large change (more than 50 nights) projected by the IPSLreg model. It is noted that IPSLglo gives smaller changes (see Table 5.1 and Part I, Sect. 3.2.1). These projected changes in very hot nights (Fig. 5.4) and very hot days (Fig. 5.3) do not take account of the urban heat island. Sensitivity modeling studies undertaken for CIRCE demonstrate that if this effect is ignored, the frequency of extreme temperature events in the future will be grossly underestimated (see Box 4.1 and McCarthy and Sanderson 2011). The urban heat island may have contributed to the extremely high temperatures experienced in Athens in summer 2007 which may provide an indication of the severity of future summers (see Box 5.1).

As well as the frequency and magnitude of extreme high temperatures, the persistence of such spells is also important in terms of human health and other impacts. Figure 5.6 shows projected summer changes in a Warm Spell Duration Index (WSDI: maximum number of consecutive days – at least six – with Tmax exceeding the 90th percentile calculated for 1961–1990) for five of the CIRCE models (all except IPSLglo). The maximum duration of warm spells increases everywhere, with the largest changes over land over the Iberian Peninsula in all models. Changes are again largest in the IPSLreg model and smallest in ENEA and CNRM. As explained in Part I, Sect. 3.4.3 the large changes over the oceans are an artifact of the lower baseline threshold values here.

For the rural case studies, the number of very cold nights (Tn5n: the number of days exceeding the 5th percentile of Tmin calculated for 1961–1990) was considered

Box 5.1 The Exceptionally Hot Summer of 2007 in Athens and Its Consequences

Summer 2007 was exceptionally hot for many areas of south-eastern Europe, the Balkan Peninsula and parts of Asia Minor with departures from the seasonal means exceeding 4°C in some areas and with distinct periods of extremely hot weather. These unprecedented warm conditions were blamed for numerous excess deaths in several countries and for serious problems in electricity supply in many regions. Greece experienced very likely the warmest summer of its instrumental history with record breaking temperatures being observed at a number of stations. Analysis of the historical air temperature record of the National Observatory of Athens (NOA), extending back to the nineteenth century, highlights the rarity of the event. Seasonal (June to August) temperature anomalies at NOA exceeded 3°C corresponding to more than three standard deviations with respect to the 1961–1990 reference period. The record value of 44.8°C was observed at NOA on 26 June 2007 (the previous record was 43°C in June 1916) during the first and most intense heat wave that affected the area. Analysis of temperature data from other less urbanized stations indicates that the urban heat effect in Athens contributed to the anomalies of the nocturnal temperatures (Founda and Giannakopoulos 2009).

The 2007 fires destroyed 280,000 ha (corresponding to about 2% of the entire Greek land surface) of forests, bush and agricultural areas, including olive groves that provide one of the main exportable Greek products. As a direct consequence of the fires, 64 human deaths occurred and thousands of people were left homeless. This was an unprecedented ecological disaster not only for Greece, but for the whole Mediterranean area.

The exceptionally hot summer of 2007 provides a strong indication of what eastern Mediterranean summers could resemble in the future (Founda and Giannakopoulos 2009).

Table 5.2 provides a comparative assessment of climate statistics for Athens, observed and simulated (nearest grid box) for 1961–1990, as well as for future climate in the middle (2021–2050) and the latter (2071–2100) part of the twenty-first century. Weighted ensemble-mean model results for maximum and minimum temperature are presented for three ENSEMBLES regional climate model (RACMO2, REMO and HadRM3) runs for the A1B emissions scenario. The weights reflect the ability of each model to simulate present-day large-scale circulation, mean climate and extremes (van der Linden and Mitchell 2009). The model statistics are in good agreement with observations, both in terms of the means and the higher statistical moments (variance and extremes). The models indicate a mean increase in summer Tmax and Tmin of about 2°C for 2021–2050 and of about 4°C for 2071–2100 compared to 1961–1990 (Table 5.2).

(continued)

Box 5.1 (continued)

Table 5.2 Statistics of summer maximum and minimum temperature, observed for the period 1961–1990 and summer 2007 (NOA), and simulated for the periods: 1961–1990, 2021–2050 and 2071–2100

	Observations (1961–1990)		Models (1961–1990)		Models (2021–2050)		Models (2071–2100)		Summer 2007 (observations)	
	Max	Min	Max	Min	Max	Min	Max	Min	Max	Min
Mean temperature (°C)	31.6	21.6	31.8	20.6	33.8	22.6	35.9	25.0	34.9	24.4
Standard dev. (°C)	3.1	2.4	3.2	2.2	3.3	2.3	3.5	2.6	3.9	3.0
95th percentile (°C)	36.3	25.4	37.2	24.5	39.2	26.4	41.9	29.1	41.5	29.2

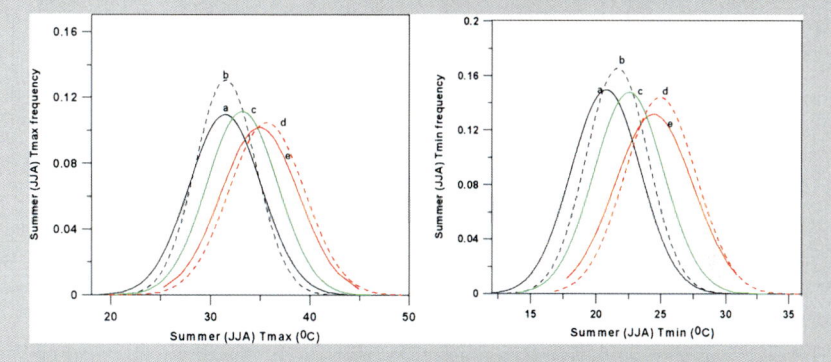

Fig. 5.5 Probability density function (pdf) of Gaussian distributions fitted to Athens JJA maximum (*left*) and minimum (*right*) temperature for the following cases: (*a – solid black*) model output for 1961–1990, (*b – dashed black*) NOA observations for 1961–1990 (*c – green*) 2021–2050 model simulations, (*d – dashed red*) 2071–2100 model simulations and (*e – solid red*) summer 2007 (NOA observations)

Figure 5.5 compares probability density functions (pdf) for different periods. The pdfs were constructed by applying Gaussian fits to summer Tmax and Tmin for 1961–1990 (both observations and the weighted ensemble-mean model results), 2021–2050, 2071–2100 and summer 2007 (observations only). It is evident that the weighted ensemble-mean pdfs for the control run are in good agreement with observations, providing a level of confidence as to the models' ability to reproduce the current climate. The pdfs for both future periods indicate not only an increase in mean but in variance as well. The observed pdf for summer 2007 lies closest to the pdf for 2071–2100, indicating that what is classified as 'exceptional' now, may be termed 'normal' or 'typical' by the end of the century (Founda and Giannakopoulos 2009). This is essentially the same conclusion that Schär et al. (2004), Beniston and Diaz (2004)

(continued)

Box 5.1 (continued)

and Beniston (2004) reached with respect to summer 2003 in Western Europe. Thus the heat waves of summer 2007 in south-eastern Europe can be viewed as further evidence, in addition to the 2003 summer for Western Europe, of exceptionally hot summers for the present period which may become typical summers by the latter part of the century, and requiring appropriate response actions by both scientists and policy makers.

For Athens, it was also estimated whether summer 2007 was an exceptional event in terms of fire risk. Using the same regional climate model output, future fire risk was compared with that for the present and 2007 to determine the severity of summer 2007 in terms of fire danger. It was found that 2007 stands out as an exceptionally extreme year with fire risk values considerably exceeding those for 2021–2050 and reaching the values projected for 2071–2100. In terms of extreme fire risk, the probability of such events was 99% in 2007, compared with only 60% for 1961–1990 and 95% for 2071–2100.

---

as an indicator (Fig. 5.7). The frequency of these events decreases by about 4 days in summer, 3 days in spring and autumn and 2 days in winter. The inter-model spread is considerably less than for the Tmax-based indicators.

Annual and/or extended winter season (October to March) total precipitation was also selected as a climate-related indicator (Fig. 5.8). The ensemble mean changes are all negative except for Alexandria and the West Nile Delta where no change is indicated. The projected decrease in precipitation is quite large for Beirut, where water shortage is already an issue (Sect. 1.4.2), as well as for Athens and the important agricultural regions of Tuscany and Apulia. The projected decrease for Apulia seems to be stronger than the very weak decrease indicated by analysis of seven of the ENSEMBLES RCMs. Although the CIRCE ensemble-mean annual change of about 50 mm is similar for Athens and Beirut, the change in percentage terms is nearly twice as large for Athens (−13% compared with −7% for Beirut). In some cases, the inter-model spread is quite large (the MPI model indicates a particularly large decrease in annual precipitation of −42% for Athens), but the magnitude of change does not vary systematically by model in the same way as occurs for temperature. Occasionally a model indicates a small increase rather than a decrease (for example, the ENEA model for extended winter season precipitation in Apulia and the MPI model for the Judean Foothills).

Changes in precipitation extremes are particularly relevant for the urban case studies. Thus Fig. 5.9 shows projected changes for these case studies in three indices calculated using daily output from the ENEA and IPSLglo models: the maximum dry spell length (consecutive dry days: CDD), heavy precipitation (the 90th percentile of daily precipitation for 1961–1990: pq90) and maximum 3-day precipitation (px3d). It is rather difficult to draw clear messages even concerning the

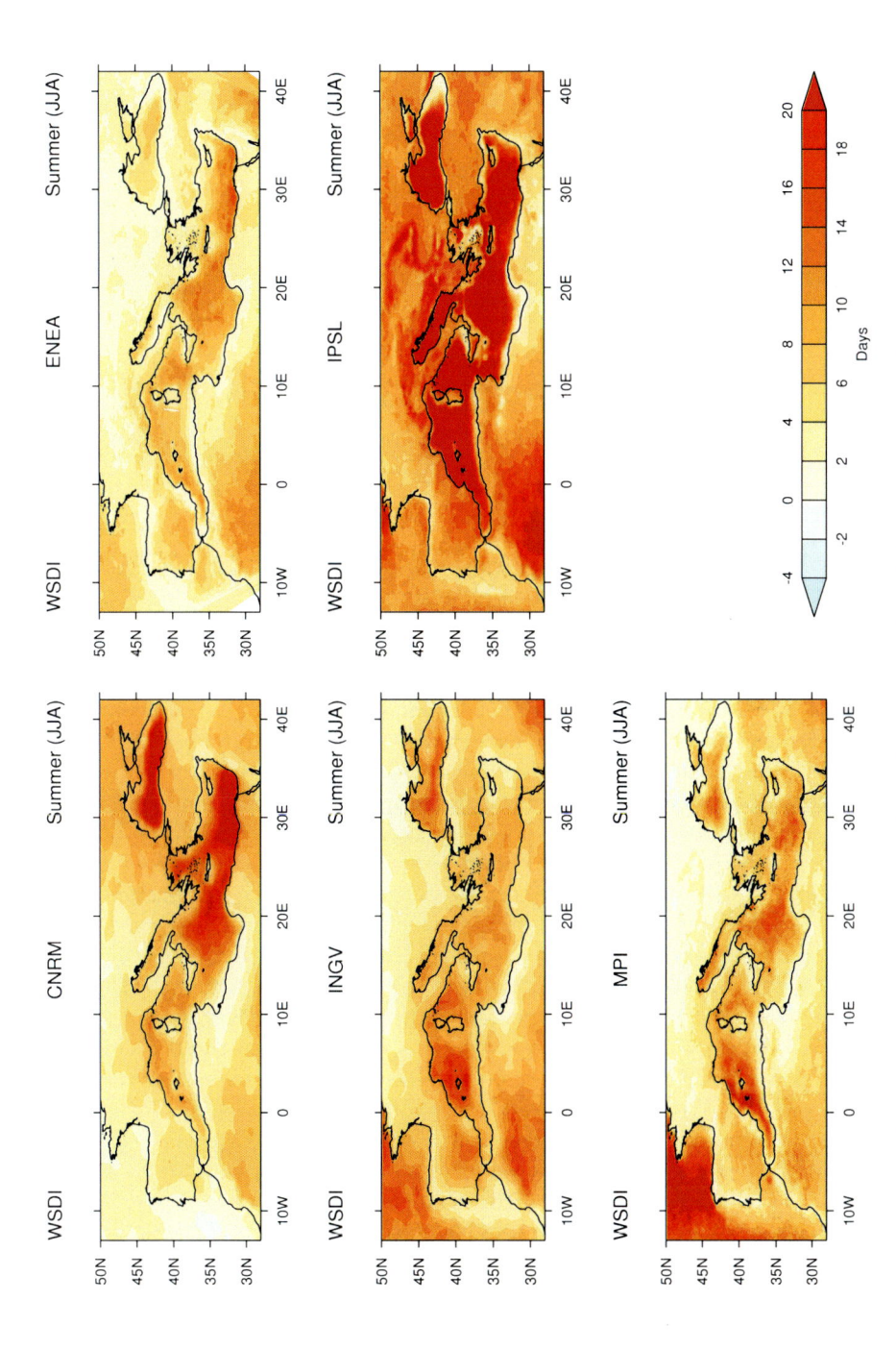

5 Integration of the Climate Impact Assessments with Future Projections 117

Fig. 5.6 Projected changes (2021–2050 minus 1961–1990) in summer WSDI (Warm Spell Duration Index: maximum number of consecutive days (at least 6 days) with Tmax >90th percentile) for five CIRCE models (CNRM, ENEA, INGV, IPSLreg, MPI)

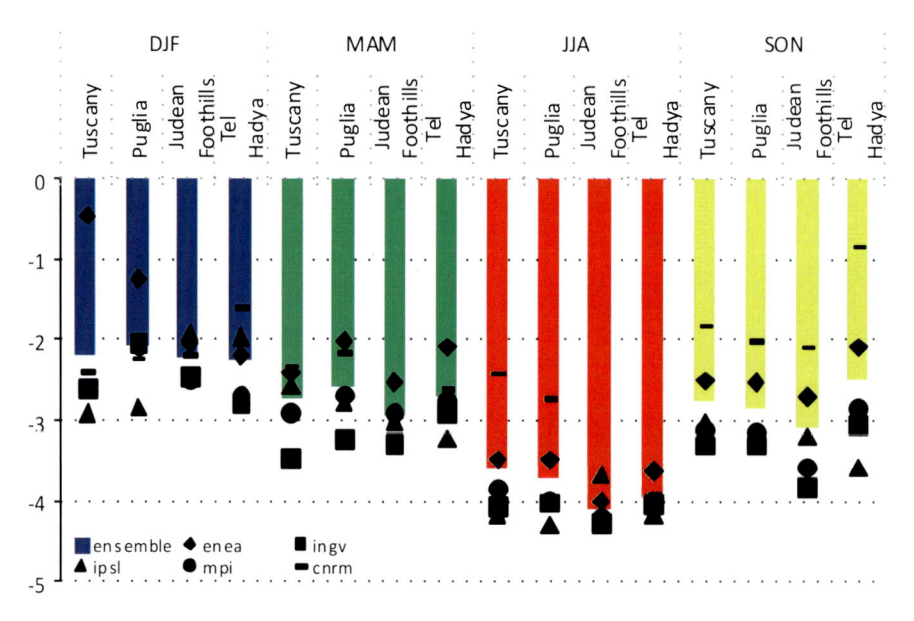

Fig. 5.7 Projected seasonal changes (2021–2050 minus 1961–1990) in the number of very cold nights (Tn5n) based on CIRCE model data for the rural case studies. The ensemble-mean change is shown, together with changes for the five individual models. IPSL is IPSLreg

direction of change in extremes from this figure. In general, the changes are rather small, particularly given the strong inter-annual variability of precipitation. Both models suggest an increase in the maximum 3-day precipitation for Athens – while for Alexandria and Beirut, ENEA indicates an increase and IPSLglo indicates a decrease (with the exception of the projected decrease for Beirut in IPSLglo, these changes are, however, rather small). With respect to CDD, both models indicate small increases for Athens and Alexandria, whereas decreases (i.e., shorter dry periods) are indicated for Beirut. These changes do not appear fully consistent with the fairly robust projections for longer dry periods across the Mediterranean, particularly during summer, emerging from other studies (Beniston et al. 2007; Giorgi and Lionello 2008; May 2008; Giannakopoulos et al. 2009). These studies tend, however, to focus on the end of the century and it may be that any signal of change in extremes has not yet emerged from the noise due to natural variability by 2021–2050. The general lack of change in precipitation extremes for Alexandria reflects the lack of change in mean precipitation seen in the CIRCE models (Fig. 5.8), and it is possible that larger changes in extremes might be apparent for the other urban case studies in the other CIRCE models. There is, for example, some indication of an increase in the frequency of heavy precipitation days over the northeastern Mediterranean in the INGV simulation (see Part I, Sect. 3.4.3). Uncertainties in projections of precipitation extremes are discussed further in Sect. 5.4.

The projected decreases in mean precipitation for the case studies (Fig. 5.8), together with possible increases in dry spell duration (see above) imply an increase in the occurrence of meteorological drought. For the rural case studies, changes in

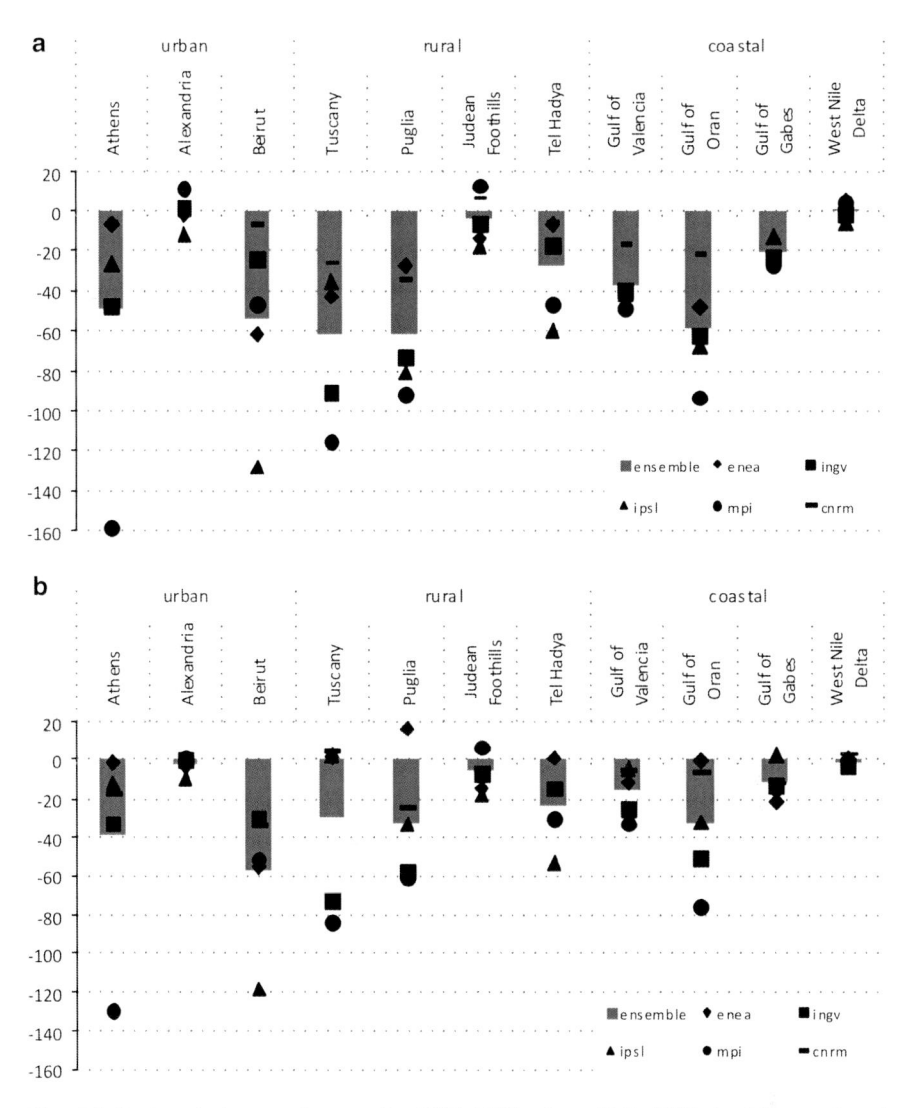

Fig. 5.8 Projected changes (in mm for 2021–2050 minus 1961–1990) in (**a**) annual and (**b**) extended winter (October to March) total precipitation based on CIRCE model data for the urban, rural and coastal case studies. The ensemble-mean change is shown, together with changes for the five individual models. IPSL is IPSLreg

agricultural drought are of concern and the Palmer Drought Severity Index (PDSI) was identified as a case-study specific indicator (Sect. 3.3.2). Changes in self-calibrated PDSI over the twentieth century have been analyzed as part of CIRCE work for the Mediterranean as a whole (Part I, Sect. 2.3.4) and for sub-domains (Sousa et al. 2011). Projected changes have not been calculated in CIRCE, but an earlier analysis of two GCM ensembles (the UK Met Office perturbed physics ensemble and the CMIP3 multi-model ensemble) indicates a mean increase of 60% (with an ensemble

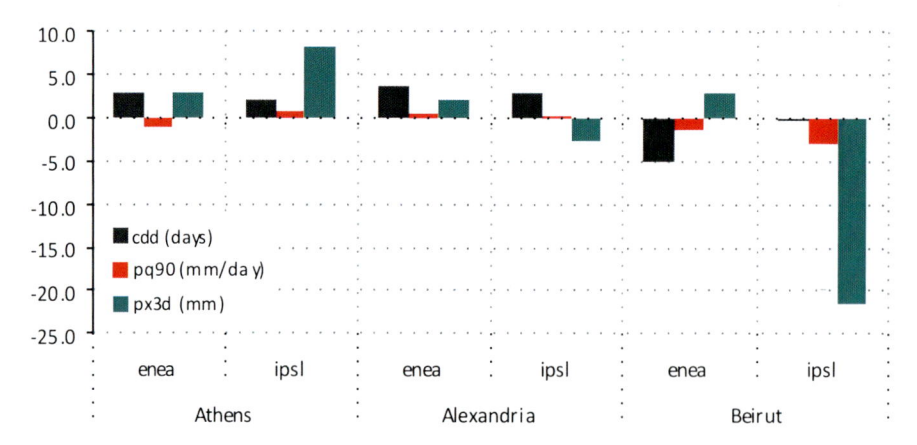

Fig. 5.9 Projected changes in three precipitation extremes indices for the urban case studies calculated using daily output from the ENEA and IPSLglo models: the maximum dry spell length (consecutive dry days: cdd), heavy precipitation (the 90th percentile of daily precipitation: pq90) and maximum 3-day precipitation (px3d). A threshold of >0.5 mm was used to define a rain day

uncertainty range of about 20–80%) in the area of the Mediterranean affected by drought defined using the PDSI and assuming a doubling of the pre-industrial CO_2 concentration (Burke and Brown 2008). In another analysis of the CMIP3 GCM ensemble, the Mediterranean is identified as one of the regions with the largest and most robust changes in short-term (4–6 months) and long-term (greater than 12 months) drought defined using soil moisture (Sheffield and Wood 2008). The latter study also indicates the year of detecting statistically significant changes in drought frequency from the baseline (1961–1990) assuming the A1B emissions scenario. For the Mediterranean and, short term drought, long-term drought and area of drought, the years of detection are 2046, 2033 and 2009 respectively. This implies that detectable changes in drought will be evident during the 2021–2050 scenario period considered by CIRCE.

The variables and indicators discussed above all relate to the land areas of the Mediterranean. For the coastal case studies, the characteristics of the Mediterranean sea surface simulated by the coupled CIRCE models are also important, in particular, sea surface temperature and salinity (see Part I, Sect. 3.3.2).

All CIRCE models simulate an increase of sea surface temperature (SST) with projected changes across the Mediterranean ranging from +0.8 to +1.8°C for the period 2021–2050 relative to 1961–1990. Sea surface salinity (SSS) mostly shows negative changes but the inter-model range is large, from +0.05 to −0.35 psu (see Part I, Table 3.4). As the water budget at the sea surface is generally projected to increase (i.e., more water loss due to a decrease in precipitation and river runoff and an increase in evaporation), this suggests a major forcing by the Atlantic freshwater input through the Strait of Gibraltar (see Part I, Sect. 3.3.4).

Figure 5.10 shows simulated mean annual SST averaged over the Gulf of Gabès from the ocean components of five of the CIRCE coupled models [INGV, LMDglo (also referred to as IPSLglo), MPI, MF (CNRM in Table 5.1) and ENEA] and two

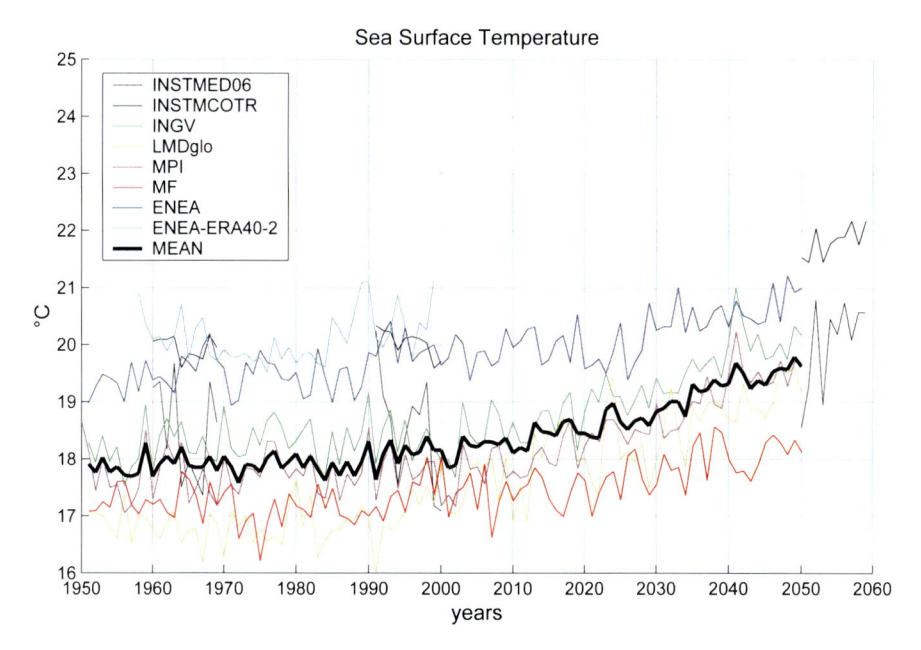

5 Integration of the Climate Impact Assessments with Future Projections 121

Fig. 5.10 Simulated sea surface temperatures for 1950–2050 in the Gulf of Gabès from CIRCE simulations. The *thick black line* is the ensemble mean

forced ocean models (INSTMED06 and INSTMCOTR – not listed in Table 5.1) together with the ensemble mean. Details of the Mediterranean Sea ocean models are given in Part I, Sects. 3.1.3 and 3.3. All models show an increase of SST over the simulation period, with acceleration of the trend during the twenty-first century (Fig. 5.10). For the period 2021–2050, the mean SST warming in the Gulf of Gabès is 0.42°C per decade, compared with 0.54°C per decade for air temperature. These values are fairly close, reflecting how fast the shallow gulf responds to the air warming. There is no statistically significant change in SSS projected for the Gulf of Gabès (not shown).

Sea-level rise is a major concern for all the CIRCE case studies and local factors are fundamental for a proper assessment of this risk. In the West Nile Delta, for example, the combination of subsidence and tectonic activity strongly influences the relative sea level with trends that are comparable, and even larger, than the projected sea-level rise (El Sayed Frihy et al. 2010). The evolution of the Valencia coast is also the result of contrasting factors: in this case, formation factors of riverine origin (such as sedimentation) and reduction factors of marine origin (such as storm surges and wind-generated waves) whose effect is, however, strongly modulated by sea-level rise at the regional scale (Sánchez-Arcilla et al. 2008).

Mediterranean sea level will increase in the future as the mass addition signal associated with ice melting penetrates from the Atlantic Ocean, but this effect will be superimposed on other regional factors, amongst which steric effects (due to volume changes) are potentially the largest. Their analysis is difficult because of simultaneous halosteric (with a more salty Mediterranean producing a contraction of its volume) and thermosteric (with a warmer Mediterranean producing an expansion of its volume) effects. The evaluation of these contrasting tendencies in climate projections is rather uncertain. An earlier assessment indicated a very large range for the end of the twenty-first century from −22 to +31 cm, with results strongly dependent on emissions scenario, model and area within the Mediterranean basin (Marcos and Tsimplis 2008).

In CIRCE, sea surface height (SSH) has been examined through the simulated steric effect (see Part I, Sect. 3.3.3). The steric effect is considered a good indicator as it constitutes a significant component of SSH variability (see above). For the period 1961–1990, the CIRCE ensemble-mean indicates a lowering of the steric effect of −0.6 cm per decade, but ranging across the models from −5.7 cm per decade to +1.7 cm per decade (see Part I, Table 3.5). The ensemble-mean trend becomes positive after 1990; with projected changes reaching +10 cm in 2021–2050 and nearly +14 cm in 2050 (relative to the period 1961–1990). If it is assumed very roughly that the steric component accounts for 70% of the total sea-level change (Bindoff et al. 2007), then the total projected sea level rise would be +14 cm for 2021–2050 and +20 cm in 2050.

A more detailed assessment of projected changes in SSH has been undertaken for the Gulf of Gabès. For the INSTMCOTR model, SSH represents the contribution of the water circulation with the central Mediterranean Sea as reference. It indicates a small positive, but statistically insignificant, trend of +0.1 cm per decade. The INSTMED06 model simulates both the halosteric and thermosteric effects and the projected changes are much higher: +1.8 cm per decade between 1960–1969 and 2050–2059 and 2.5 cm per decade between 1991–2000 and 2050–2059.

In addition to changes in mean sea level, potential changes in extreme sea level and storminess also need to be considered. Studies of marine storminess for the whole Mediterranean basin suggest that the mean significant wave height (SWH) field over a large part of the Mediterranean Sea will be lower in future scenarios than in present conditions (Lionello et al. 2008). There are, however, a few exceptions: in particular, two areas between Greece and Northern Africa and between Spain and Algeria, where SWH is projected to be higher. Analogously, extreme SWH values are smaller in future scenarios than in the present climate except for the presence of higher SWH extremes in the central Mediterranean during summer for the A2 emissions scenario. This is substantially consistent with the analysis of storm tracks, which show reduction during late summer and autumn over the Mediterranean region in future scenarios (Lionello et al. 2008). Dedicated studies of storm surges have been carried out mainly for the Northern Adriatic Sea and do not identify substantial changes in frequency of high storm surge events (Lionello et al. 2003, 2010). As part of CIRCE, the WAM ocean-wave prediction model (WAMDI 1988) was driven by wind fields from the MPI CIRCE simulation. The results suggest that the future winter wave climate will be characterized by a shift of the whole distribution towards lower values – with a decrease in mean and extreme (both high and low) values.

In conclusion, there is no evidence of increased risks posed in the Mediterranean coastal zone by changes of marine storminess except for those associated with the large uncertainty in future mean sea level (see above and Sect. 5.4), which makes it difficult to reach robust conclusions. Although the Mediterranean Sea will on multi-decadal timescales follow the projected large increase in global sea level, the actual regional values may be substantially different from the global one.

The material presented in this section demonstrates that it has been possible to produce quantitative projections for many of the climate hazards identified for the urban (Fig. 4.2), rural (Fig. 4.6) and coastal (Fig. 4.11) case studies (see also Table 5.3). The underlying CIRCE and ENSEMBLES climate model simulations are, however, subject to various model biases including a systematic cold bias in the CIRCE models. These and other sources of uncertainty associated with the climate projections are discussed in Sect. 5.4.

5.3 Biogeophysical and Socioeconomic Projections

5.3.1 Introduction

As part of the CIRCE case-study work, a subset of the potential biogeophysical and social indicators (Chaps. 3 and 4) was identified for which it has been (or should be) possible to provide quantitative projections either through process-based or statistical modeling (Table 5.3). These include a fire weather index/forest fires (AT, TU); ozone exceedance/anomaly (AT, AL); all-cause mortality (AT, BE, AL, OR); energy production/consumption (AT, VA); marine characteristics such as, salt-water intrusion (VA), marine pollution (OR), invasive species (GA), and vulnerability to sea-level rise (WN); tourism (GA, TU, AP); and wheat yield (all the rural case studies). It is also possible to provide quantitative estimates of some vulnerability indicators that are based on climate variables, such as a temperature-defined tourist season, or a rainfall and temperature-defined drought indicator. For other indicators (Table 5.3), there is a lack of appropriate data and/or models to allow quantitative assessments of the consequences of climate change, e.g., for marine water quality in Valencia, shoreline changes in the Gulf of Gabès, and food-web changes in the Gulf of Oran. For these indicators, there is nonetheless, sufficient local evidence and knowledge of current trends and response to climate and other system pressures to allow more qualitative assessments of climate change impacts.

There are considerable challenges in being a down-stream user in projects like CIRCE. There are inevitable time delays in the availability of climate model runs and further necessary time lags for post-processing climate model data for impacts modeling. Within the time-limited framework of writing this volume, it has been possible to assess the potential consequences of climate change for two case studies in more detail than for the others: a coastal case study, the Gulf of Gabès

Table 5.3 Capacity for quantitative or qualitative assessments of the consequences of future climate change for the CIRCE integrated case studies

Case study[a]	Quantitative	Qualitative
AT:	Peri-urban forest fire frequency and surface area burnt (I&V)	Water for cooling plants (V)
	Fire Weather Index (number of days >15; number of days >30) (V)	Area of peri-urban forest (V)
	All-cause daily mortality (I)	Vehicles per km^2 (V)
	Energy consumption – domestic and commercial (I)	Building density (V)
	Ozone exceedance days (V)	Air conditioning use (V)
	Greece urban population growth rate (V)	Investment in fire prevention/protection (V)
	Athens population (V)	Access to health care (V)
	Gross Domestic Product (per capita; current prices) (V)	% population with pre existing disease (V)
	Life expectancy at birth, Greece (V)	
BE:	**All-cause mortality** (I)	**Saline intrusion in wells** (V)
	Waterborne and food-borne disease: total, dysentery, typhoid, hepatitis A (I)	Drinking water quality (V)
	Health damage due to water scarcity (I)	Access to clean water per capita (V)
	Annual cost of waterborne disease (V)	**Water supply capacity** (V)
	Population growth (V)	Groundwater abstraction (V)
		Cost of desalination (V)
		Water resource management (V)
AL/WN:	Total Ozone column anomaly (V&I)	Fire pixel count (V&I)
	All-cause mortality (I)	Aerosol types (dust, pollution, mixed, clean) (V&I)
	Coastal inundation (I)	Saltwater intrusion (V)
	Vulnerability to sea level rise by land-use type (V)	Water use by sector/capita (V)
	GDP per capita (V)	Industrial output (V)
	Population growth (V)	Marine water quality (V)
		Level of environmental protection / management (V)
TU:	**Rain-fed wheat yield** (I)	**Land use** (V)
	Domestic tourist arrivals (I)	Traditional Tuscan rural landscape (V)
	Fire Weather Index (V)	**Water consumption by sector** (V)
	Fire frequency (I)	Salinization (V)
	Water availability (V)	Water resource management (V)
		Energy demand (V)
		Gross Domestic Product (V)
AP:	**Wine, wheat, olive oil crop yield** (I)	Land cover class (V)
	Annual number of tourists (I)	Water rationing (V)
	Wooded area burnt (I)	Sector contribution to GDP (V)
	Potential evapotranspiration (V)	% Employed in agriculture, tourism (V)

(continued)

5 Integration of the Climate Impact Assessments with Future Projections 125

Table 5.3 (continued)

Case study[a]	Quantitative	Qualitative
	Environmental water budget (V)	Salinization (V)
	Water consumption per capita (V)	
	Seasonal Shift Index (V)	
	Population growth (V)	
TH:	**Rain-fed wheat yield** (I)	Water availability (V)
	Wheat yield and above ground biomass in response to supplemental irrigation (adaptation)	Salinization (V)
	Groundwater levels (V)	Water resource management (V)
	GDP per capita (V)	Area of arable land (V)
	Population growth (V)	% Employed in agriculture (V)
JF:	Grain yield in rain-fed wheat (I)	Land use (V)
	Forest tree growth (stem volume) (I)	Proportion of protected land (V)
	Natural vegetation types (I)	Tree health survey (V)
	Tourist visits; domestic and international (I)	Water resource management (V)
	Soil respiration/Carbon sequestration (V)	Soil quality / erosion (V)
	Population growth in Israel (V)	Pests and diseases (V)
	Urban / Rural population in Israel (V)	Critical species thresholds (V)
	Gross Domestic Product per capita in Israel (V)	% Employed in agriculture / tourism (V)
	Employment/unemployment rate in Israel (V)	
	Population density – Judean Foothills (V)	
	Water production/consumption by sector (V)	
VA:	Salt-water intrusion (I&V)	**Marine water quality** (V)
	Shoreline erosion (I&V)	Area of land used as a coastal buffer (V)
	Electricity generation and **consumption** (I)	
	Insurance claims for property damage (I)	Area of land / population vulnerable to coastal flooding (V)
	Mortality max temperature thresholds (V)	Coastal outflow of freshwater nutrients (V)
	Population growth (V)	Pollution discharges (V)
OR:	Marine pollution discharges (I&V)	Invasive marine species (jellyfish, green algae) (I)
	Water quality parameters (riverine, estuarine sites) (I&V)	Fish stocks (I)
	All-cause mortality (I)	Food-web changes – dolphins (V)
		Ecological monitoring of vulnerable littoral sites (V)
	Grain yield (I)	**Land cover class for Oran Bay** (V)
	Runoff (V)	**Carbon uptake by land cover class** (V)
	Population of Oran (V)	Coastal protection / management (V)
	GDP per capita (V)	% employed in fishing industry (V)

(continued)

Table 5.3 (continued)

Case study[a]	Quantitative	Qualitative
GA:	**Alien marine species** (I&V)	Shoreline modification (I)
	Fish Production (I)	Coastal Vulnerability Index (V)
	Daily Climate Tourism Potential Index (DCTPI) (I)	Area of land used as a coastal buffer (V)
	Socio-economic impacts on tourism activity due to changes in the DCTPI and beach erosion (I)	Coastal management / protection (V)
	Maximum swell height (V)	Marine pollution / discharges (V)
	Seasonal shift index for tourism (V)	Salinization (V)
	Annual population growth rate (V)	% employed in tourism / fishing (V)
	Urbanization (urban, rural population) (V)	
	Gross Domestic Product based on purchasing power parity per capita GDP (V)	
	Life expectancy at birth (V)	

[a]See Appendix 1 for case-study acronyms
'I' denotes impact indicators
'V' denotes vulnerability indicators
Bold denotes those indicators which have been quantified based on climate model projections at the time of writing

(Tunisia) and an urban case study, Athens (Greece). Although this provides a somewhat limited future perspective, the chosen set of indicators derives from diverse regions (a large capital city in the European Mediterranean and a coastal semi-arid region of North Africa) and covers diverse sectors (from terrestrial and marine ecosystems to tourism, energy and human health). More detailed future information will be available for other case-study regions by the end of the CIRCE project, and will be made available from the CIRCE integrating case-studies website, http://www.cru.uea.ac.uk/projects/circe/ and also written-up in journal papers. It is anticipated, for example, that papers will be produced on projected crop-yield changes for Tuscany (durum wheat), Apulia (wheat, wine and olive oil) and Syria (durum wheat).

Each integrated assessment of the possible response to climate change is grounded in the conceptual framework of the CCIF (Chap. 1), and is primarily based on statistical impacts models but also considers more qualitative but nonetheless valuable evidence such as stakeholder knowledge of the system response to past climate events (Sect. 6.3). Future change in non-climate drivers such as population projections and, where available, future narratives for socio-economic and technology trends, are considered alongside climate projections.

In the next section (Sect. 5.3.2), relevant CIRCE Mediterranean-wide work on future changes in biogeophysical and socioeconomic systems is outlined, while Sect. 5.3.3 presents specific examples of indicators constructed for Athens and the Gulf of Gabès. Finally, Sect. 5.3.4 discusses key projected changes for all case studies in the context of the CCIF.

5.3.2 Key Messages on Mediterranean Changes in Biogeophysical and Socioeconomic Systems from the CIRCE Project

Previous parts of the RACCM report potential consequences of climate change for the Mediterranean region which are largely focused on particular sectors and at a national or wider spatial scale of analysis. The key CIRCE themes are *Water* (Part II), *Agriculture Forest and Ecosystem Services* (Part III), and *People* (Part IV). The following sub-sections summarize the key messages with respect to the future emerging from the wider CIRCE project that have particular relevance to the local and regional case studies. In many cases, this 'broad-brush' summary is based on reviews of earlier work and on earlier climate model runs rather than on impacts modeling using output from the CIRCE climate models runs discussed in Sect. 5.2.

5.3.2.1 Water Resources

Water scarcity is an issue for the entire Mediterranean region; even today 30 million people in the southern and eastern parts of the Mediterranean region do not have access to 'safe' water. Changes in the hydrological cycle of the Mediterranean could have wide-ranging implications from agricultural production through to natural ecosystems, energy production, tourism and, health and well being. Future inequalities in water supply and availability could result from complex interactions between climate system drivers and increasing pressures from socio-economic systems. The CIRCE climate models indicate drier conditions over the Mediterranean in the future (see summary of precipitation changes in Sect. 5.2; Fig. 5.8). The projected decline in evapotranspiration is strongest in summer when land-surface aridity is highest (Part II, Chap. 8).

A general decline in river discharge has been projected across the Mediterranean region by the end of the twenty-first century for the A1B emissions scenario using the super high-resolution (20 km) Japanese Meteorological Research Institute (MRI) GCM and a river-flow model also developed by MRI (Part II, Sect. 8.2.3.4, Fig. 8.9, Kitoh et al. 2008). Of particular interest to the case studies, is a CIRCE analysis of the MRI simulations which indicates a 46% decrease in annual discharge for the River Ebro (located within the Valencia-Catalan case-study region, Spain), and in contrast to this a projected increase in River Nile (Egypt) discharge of about 2,900 m^{3-s}. However, the model substantially underestimates observed Ebro river flow, and in the case of the Nile, human influence such as the building of dams is not considered.

5.3.2.2 Mediterranean Agriculture, Forest and Ecosystem Services

Based on long-term analysis of satellite records for North Africa and the Middle East, observed degradation of vegetation is particularly severe in coastal areas of

Algeria, and is moderate in large parts of the West Nile Delta and Israel. The pattern is mixed in Tunisia, Syria and Lebanon, with some areas showing moderate to advanced increases in vegetation and other areas showing moderate to severe degradation (Part III, Fig. 2.4). Recent observations suggest that Mediterranean tree species are disproportionately sensitive to climate change compared to temperate tree species, and that future climate change could further accentuate historic climate and non-climate stresses (Part III, Chap. 5).

Earlier modeling studies reported in Part III, Chap. 2 suggest a northward shift in the margins of the evergreen closed-canopy forests in the Mediterranean region, replaced with low-cover shrublands and grasslands in southern areas. As a general rule it is considered that, in response to temperature alone, the northward migration of crop zones could be in the order of 150 km for every 1°C increase, with southern and eastern Mediterranean regions most vulnerable (Part III, Chap. 2). An increase in future droughts could reduce plant biomass, increase mortality rates, and lower the capacity of the vegetation to act as a carbon sink. There is likely to be large variation within the Mediterranean region (and at the local scale, the balance of processes can be particularly complex – see Box 4.2), from a projected increase in the carbon sink in the northern boundary of the region to a reduction in carbon sequestration in southern and eastern semi-arid shrubland regions (Part III, Chap. 2). An increase in fire risk due to increased fuel dryness and reduced humidity is likely to have a greater impact in areas with a large percentage of forest land cover. There are also potential interactions between future fire risk and temperature, with warmer drier conditions slowing growth rates and recovery (Part III, Chap. 3).

Using the ENEA regional climate model CIRCE runs (see Table 5.1) coupled with process-based crop models, the CIRCE project has assessed the agricultural impacts of climate change for two future periods (2011–2030 and 2031–2050). The results (presented in Part III, Chap. 4) suggest a general advance (i.e., earlier in the year) in phenological stage and a reduced inter-phase period. Despite a shorter time for biomass accumulation, durum wheat may benefit from enhanced CO_2 fertilization effects and increases in productivity. Grapevine may be particularly vulnerable to warmer and drier conditions in the future, with yield declining in many parts of the Mediterranean (except some areas of western Asia and south-western France). In addition, projected increases in temperature imply a northward and eastward shift in the land suitable for olive tree cultivation within the Mediterranean region.

Future climate change could result in more frequent and more severe fires and a longer fire season in the Mediterranean (Part III, Chap. 6). Trends towards increased land abandonment in some parts of the region could increase fuel load and landscape continuity. Future fire vulnerability for the end of the twenty-first century has been estimated in the CIRCE project using IPCC 2007 projections for wildfires and drought occurrence. Projections show the greatest vulnerability exists in inland regions of Greece and Spain (Part III, Fig. 6.4). There is uncertainty about the net direction of post-fire feedback processes, i.e., a reduction in fuel accumulation (biomass) could lead to negative feedback in terms of reduced fire risk, but post-fire communities dominated by seeders could have increased flammability and act as a positive feedback driving further change (Part III, Sect. 6.1). There are also uncertainties with regard to the future role played by land management in fire regimes.

5.3.2.3 Mediterranean Communities

Economic Impacts

For the A1B SRES scenario, and using a Computable General Equilibrium (CGE) model based on sector impacts, the CIRCE research line, *Economic Impacts of Climate Change* (Part IV, Sect. 10.5), has estimated the direct economic impacts of climate change by 2050. At an aggregated level (for energy demand, land loss due to sea-level rise, and tourism), the economies of countries of the North African and Eastern Mediterranean are indicated to be some of the most adversely impacted (a loss of up to 2% of GDP by 2050), while the European Mediterranean countries have a lower economic vulnerability (losses of about 0.2–1.4% of GDP by 2050). For the individual case-study countries in common with this assessment (Spain, Greece, Italy and Tunisia), Tunisia is estimated to have the highest net loss of GDP (about 1.1%) and Greece the lowest net loss (about 0.4%). However, the Eastern Mediterranean region (not reported at a national level) and covering the case-study countries of Syria, Lebanon and Israel, is estimated to suffer a loss of 1.6% in GDP, emphasizing their greater economic vulnerability to climate change (Part IV, Chap. 10, Fig. 10.4).

For disaggregated sectors and using A1B SRES scenarios, percentages changes are estimated by 2050 (Part IV, Chap. 10, Table 10.12). Household electricity demand increases by about 0.7% and the demand for natural gas and oil decreases by about 12.5 and 15% respectively, with little variation among the Mediterranean countries considered (for which the CIRCE case studies include Italy, Spain, Greece, Tunisia, 'Rest of North Africa', and 'Rest of Middle East'). The proportion of total land loss due to sea-level rise ranges from −0.02% for Spain to −0.25% for Tunisia. Changes in the demand for tourism and recreational services are estimated to range from −1.3% for Italy to −4.9% for Tunisia, with changes in tourism expenditure ranging from −$1.21 billion for Tunisia to −$9.87 billion for Italy. It should be noted that the climate-change information employed in the underlying impacts modeling does not have high temporal or spatial resolution – in the tourism model, for example, climate is represented only by mean annual temperature. Moreover these assessments do not adequately account for changes in the frequency and magnitude of extreme events. While existing economic impact models cannot cope with more explicit climate data, future work within the EU (such as DG Climate: 'Climate Proofing EU Policies'), aims to incorporate existing socioeconomic scenario data as a solution to this limitation, and as part of an EU Adaptation Strategy (Swart and Biesbroek 2009).

Human Health and Well Being

Research in CIRCE (Part IV, Chap. 13) focused on present-day risks of excessive heat and air pollution, and infectious disease for ten cities of the Mediterranean region. Seven of the cities (Athens, Barcelona, Bari, Rome, Tel Aviv, Tunis and Valencia) are located within the home nations of the integrated case-studies, and one city (Athens) is an urban case study. Heat thresholds (above which mortality increases) were defined using maximum apparent temperature (which is based on air temperature and dew point) and varied across the cities, from 26°C in Barcelona to 33.6°C in Tunis. A 1% increase in apparent temperature above the threshold was found to increase heat-related deaths more on 'high' ozone days than on 'low' ozone days in Athens, Rome, and Valencia, but not in Barcelona, Tel Aviv or Tunis (Part IV, Sect. 13.3.3).

Energy

CIRCE work on energy focused on potential renewable energy resources (i.e., CSP – Concentrated Solar Power generation) within North African countries and options for greenhouse gas mitigation (Part IV, Chap. 14). There is a large potential for increase in energy consumption per capita in the countries of North Africa in keeping with population and economic growth projections. CIRCE projections for economic growth made using the general equilibrium model GEM-E3-MED CGE as part of the energy work (Part IV, Chap. 14, Table 14.3) are lower for European Mediterranean countries (annual % GDP increase of 2.4% for Spain, 1.9% for Greece, 1.6% for Italy for the period 2020–2030) than for North African countries (annual % GDP increase of 5.2% for Egypt, 4.5% for Tunisia, 4% for Algeria and Libya for the period 2020–2030). In addition to the opportunities of economic growth, there are also considered to be large opportunities for improving energy efficiencies (Part IV, Fig. 14.3) in North African countries, particularly for industry (and especially in Egypt) and household appliances (especially in Algeria). Oil is likely to dominate the energy mix in the short-term, but could be overtaken by gas around 2020. All North Africa countries also have a strong potential for renewable energy, largely in wind and solar energy, but also in geothermal energy (Part IV, Sect. 14.2).

Mediterranean Tourism

Within the CIRCE project (Part IV, Sect. 15.4) the SRES A1B scenario was used with the Hamburg Tourism Model (Bigano et al. 2008) to project changes (the difference between including and not including climate change scenarios) in national tourism arrivals by the year 2100. Generally, the inclusion of climate change scenarios (not explicitly defined) could encourage additional tourist flows to northern European countries while discouraging tourist flows from northern to southern countries of the Mediterranean. For the integrated case studies, these national results (Part IV, Table 15.3) suggest a large decline by 2100 in international arrivals for West Asia and North African countries (−52, −45, −38, −36, −32 and −25% for Algeria, Egypt, Tunisia, Israel, Syria, and Lebanon respectively), and a lower decline in international arrivals for European Mediterranean countries (−22% for Greece, and −15% for Spain and Italy). The projected percentage changes are lower for domestic tourism (ranging from −17% for Algeria to +10.5% for Italy) and expenditure (ranging from −27% for Tunisia to +5% for Italy).

In a local vulnerability assessment of tourism in the north-eastern coastline of the island of Djerba, Gulf of Gabès (Duvat and Magnan 2009), Tunisia, a survey of decision makers at tourist resorts was used to assess local vulnerability in terms of sensitivity (the presence of dunes and a terrestrial/marine buffer zone) and exposure (e.g., the location of the site relative to waves and storms, distance from sea and level of protection for infrastructure). The tourist resorts on the northern stretch of coastline (Part IV, Fig. 15.3) were found to score more highly in terms of physical environmental vulnerability. This area has been subjected to rapid sand loss and hard engineering coastal structures have accentuated the environmental sensitivity (Part IV, Sect. 15.5.2).

5.3.3 Specific Examples from the CIRCE Case Studies

Two example case-studies are presented in this section, to illustrate the potential for the construction of quantitative indicators of climate change: first, for the urban case study Athens (Greece) focusing on peri-urban fires, air pollution, human health and energy; and second, for the coastal case study in the Gulf of Gabès (Tunisia) which focuses on marine ecosystems and coastal tourism. In each case, the estimates of future hazard, impacts or vulnerability are based on climate model projections.

5.3.3.1 Athens Urban Case Study

For the Athens urban case study, biogeophysical and vulnerability indicators based on system thresholds (Table 4.1) were initially calculated using regional climate model output from the earlier EU ENSEMBLES project (van der Linden and Mitchell 2009) and, as it became available, output from some of the CIRCE models. The ENSEMBLES models were run over Europe at a resolution of about 25 km, for the A1B emissions scenario and the period 1950–2050 or 2100. Unlike the CIRCE model runs (Sect. 5.2) these are not coupled atmosphere–ocean regional models.

The response of biogeophysical and social systems to past extremes in climate can be used as an analogue for the potential consequences of climate change in a region. Thus for example, the Athens case study also assessed the impacts of the exceptionally hot and dry summer of 2007 on fire risk (Box 5.1).

Peri-Urban Fires

Greece as a whole recorded an average of 56,000 ha of forest burned per year between 1990 and 2000, with most of these caused by human ignition (Goldammer and Mutch 2001). The destruction of forests by fire is of great concern since it can cause many negative side effects including floods, soil erosion and consequent loss of fertility. Furthermore, peri-urban forest fires can play a role in regulating air

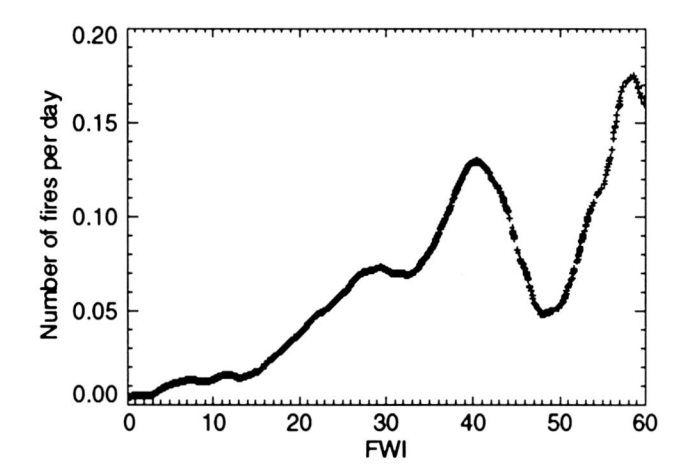

Fig. 5.11 Mean number of fires per day against FWI (*crosses*) for fires near Athens for the period, 1983–1990. The crosses merge to form a *thick curve*, except at large FWI where data are sparse (Good et al. 2008)

temperature and wind circulation in the surrounding city (Court-Picon et al. 2004; Iliadis 2004). Thus peri-urban forest fires may contribute to an increase of temperature in the city during the summer months and an intensification of the urban heat island (Kurz et al. 1995).

Peri-urban forest fires are highly sensitive to climate change since fuel moisture is affected by precipitation, relative humidity, air temperature and wind speed. For the Athens case study, fire risk for Attica and the surrounding forest area was assessed using the Canadian Fire Weather Index (FWI, Van Wagner 1987). This is a numerical rating of fire intensity calculated using daily maximum temperature, relative humidity, wind and precipitation. Although the FWI was originally developed for Canadian forests, several studies have shown its suitability for the Mediterranean basin (e.g., Viegas et al. 1999).

In order to identify a relationship between FWI values and fire occurrence for the Attica region, meteorological data for the period 1983–1990 were obtained from the Greek National Meteorological Service and daily fire frequency and area burnt in the regions of the meteorological stations were obtained from the Forest Research Institute of Athens. The locations of the fires are not available, although all were within about 150 km of the meteorological station. Fire occurrence was found to be low for a FWI less than about 15 and tends to increase with increasing FWI (Fig. 5.11). The oscillations at high FWI (above FWI=45) may be due to sparse data, but could be an artifact of human behavior: during heatwaves, human activity is reduced and so fewer fires may be started (Venevsky et al. 2002). A threshold of FWI=30 was found to best define extreme fire risk in the peri-urban region of Attica.

Future fire risk was then estimated using daily output from six ENSEMBLES RCM (KNMI-RAMCO2, CNRM, ETHZ, MPI, METO and METNO) simulations.

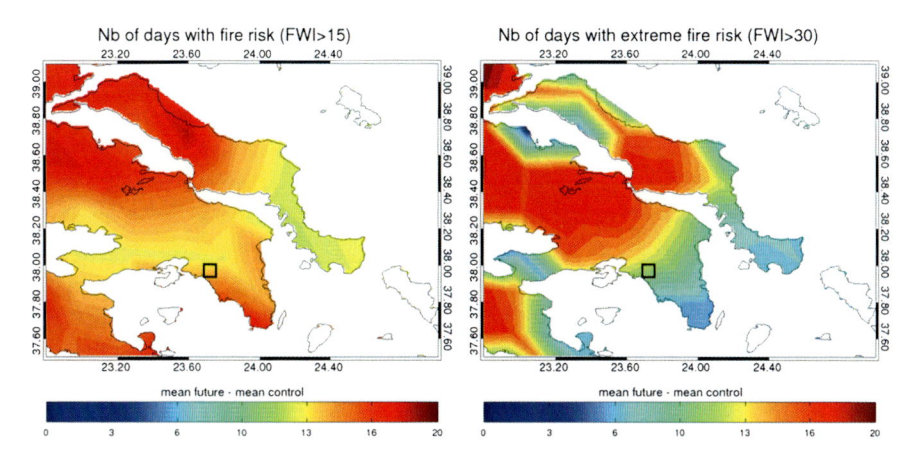

Fig. 5.12 Projected changes (2021–2050 minus 1961–1990) in the number of days with fire risk FWI > 15 (*left*) and extreme fire risk FWI > 30 (*right*) based on output from six ENSEMBLES RCM simulations

Ensemble-mean changes (with respect to the 1961–1990 baseline) for two 30-year future periods (2021–2050 and 2071–2100) were calculated. For Attica and the surrounding forest areas, the future fire risk is projected to increase in both periods. For the near-future period 2021–2050, the number of days with fire risk (FWI > 15) increases by up to 17–20 days or more in the north-western parts of the study area, while extreme fire risk (FWI > 30) is more spatially variable, with an increase of about 10 days in the vicinity of Athens itself (Fig. 5.12). By the end of the century (2071–2100), the increase is 50 days (for FWI > 15) and 40 days for extreme fire risk (not shown). The same analysis based on two of the CIRCE RCM simulations (ENEA and MPI) shows a different spatial distribution for FWI (Fig. 5.13). In peri-urban regions, the future fire risk for FWI > 15 is greater in the Greater Athens Area than in the ENSEMBLES simulations. In contrast, extreme fire risk shows smaller increases in the CIRCE simulations (Fig. 5.13) compared to the ENSEMBLES simulations (Fig. 5.12).

Air-Pollution

Climate change is expected to affect air quality. It is already known that day-to-day meteorology affects both the severity and duration of pollution episodes. In a future climate-change world, faster chemical reactions, increased biogenic emissions, and stagnation may contribute to an increase in the occurrence of ozone pollution episodes. Ozone episodes are closely linked to adverse effects on human health, vegetation and ecosystems, and there is a documented relationship between ozone exceedances and hospitalization.

It has previously been shown that daily maximum temperature is strongly related to ozone concentration and accounts for much of the influence of meteorological

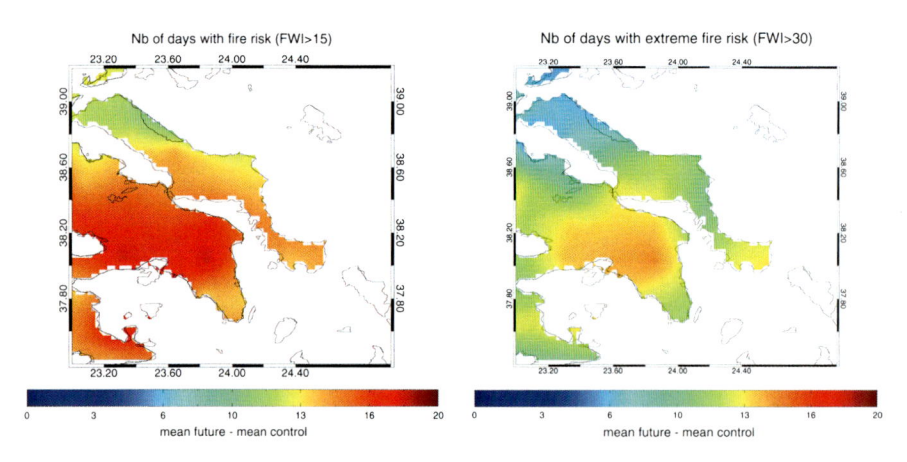

Fig. 5.13 Projected changes (2021–2050 minus 1961–1990) in the number of days with fire risk FWI > 15 (*left*) and extreme fire risk FWI > 30 (*right*), based on the average of two CIRCE models (ENEA and MPI)

variables on ozone (Lin et al. 2001; Bloomer et al. 2009). Moreover, the strong correlation of ozone pollution episodes with temperature is associated primarily with surface air ventilation, since high temperatures are typically a result of stagnant air. Secondly, the correlation of ozone with temperature also depends on local ozone production chemistry and on temperature-dependent biogenic emissions (Jacob et al. 1993).

The relationship between daily 8-h maximum temperature and ozone concentration for Athens is shown in Fig. 5.14. The probability of ozone exceedance days (defined as days with maximum 8-h average 60 ppb, i.e., according to EU Directive 2008/50/EC) increases over the temperature range 17–38°C, mainly due to high activity of photo-chemical processes.

The statistical relationship shown in Fig. 5.14 (the green line) was applied to output from three CIRCE models (ENEA, MPI, and IPSLglo) to estimate the potential impact of increasing future temperatures on ozone exceedance days in the Greater Athens Area. Figure 5.15 indicates that the number of these days is projected to increase. For the future period 2021–2050, ozone exceedance days are expected to increase by 4 days per year for MPI, 8 days for ENEA and 34 days for IPSLglo. A previous analysis (not shown) using output from the KNMI ENSEMBLES model found an increase, amounting to about seven extra ozone exceedance days per year. For the end of the twenty-first century (2071–2100), the KNMI simulation suggests an increase of almost an additional month (about 27 days) of ozone exceedance days per year.

Published studies of climate change impact on air quality at the regional scale in Europe indicate worsening of future air-quality and especially increased occurrence of ozone exceedance days. These results have been obtained by coupling process-based climate-chemical models at various horizontal resolutions and for various greenhouse gas emission scenarios. Forkel and Knoche (2007), for example, using

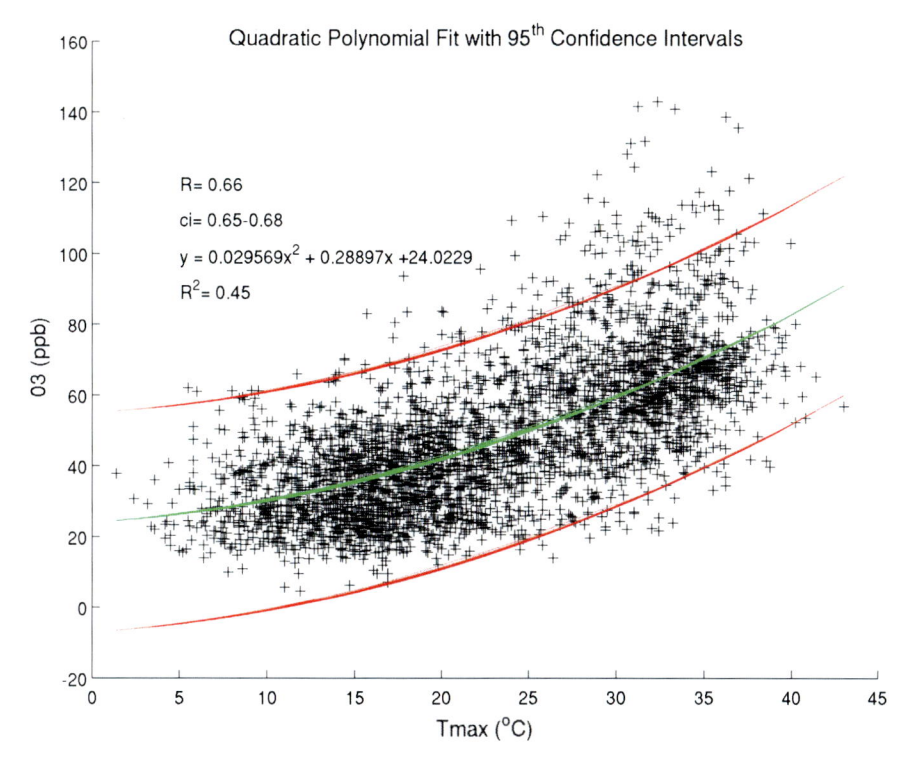

Fig. 5.14 Scatter plot of the daily 8-h maximum average ozone concentration vs. daily maximum temperature for Athens, with best fit line (*green*) and 95% confidence limits (*red*)

a varied-resolution climate-chemical model system under the IS92a emission scenario, found that the impact of climate change on its own could lead to an increase of about 16 extra ozone exceedance days in Central Europe by the 2030s compared to the 1990s. Results from such dynamical modeling studies provide qualitative support for the statistical modeling results presented here.

Human Health Risks

In order to explore future changes in heat-related deaths, all-cause daily mortality data for 1992–2006 for Athens were acquired from the Greek National Institute of Statistics. Daily climate data were provided by the National Observatory of Athens. These data show that there have been considerable heat-related deaths in Athens, both from moderate and extreme heat during summer months due to individual heat-wave events. A clear seasonal variation in mortality is observed with more deaths in winter and in summer, and fewer deaths during the transitional seasons.

An empirical-statistical model for summer heat stress was constructed using these data and applied under the assumption that the relationship between temperature

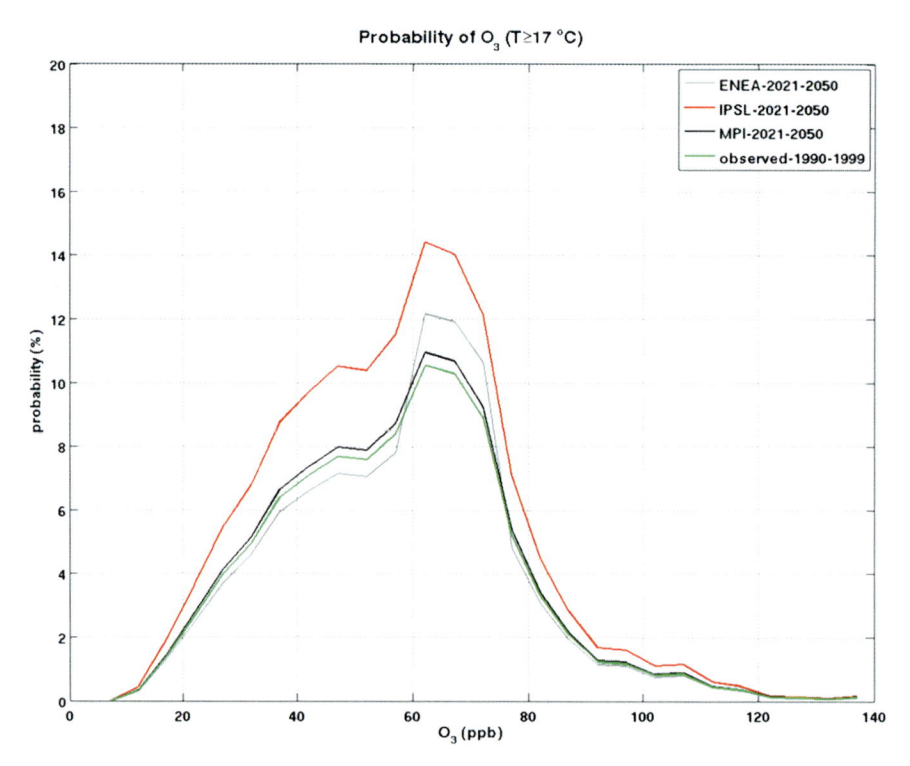

Fig. 5.15 Ozone probability distributions for temperature above 17°C, for two periods, observed (1990–1999) (*green*) and 2021–2050 (*black, red, grey*) for CIRCE models (ENEA, MPI, IPSLglo). Values are plotted at the mid-point concentration of each 5 ppb ozone concentration bin

and mortality is linear. Excess deaths were calculated as deaths above the expected mean daily value for each summer month in the observed period (78.9 deaths in June, 81.2 in July and 79.1 in August). Thus heat-related deaths are defined as the number of deaths occurring in excess of the number that would be expected for the population in the absence of stressful weather.

Summer daily excess deaths in Athens for each maximum air temperature interval are presented in Fig. 3.8 along with the frequency of occurrence of the temperature intervals. A 'V-shaped' relationship between mortality and temperature is evident. Hotter days are associated with greater mortality risk above about 34°C (which is therefore used as a system threshold – see Table 4.1). Substantial heat-related deaths occurred at very high temperatures.

A linear model was then used to predict heat-related mortality for future climate (2021–2050 and 2071–2100) using temperature output from the RACMO2 RCM ENSEMBLES simulation for the A1B emissions scenario (Fig. 5.16). An 'adaptation' or acclimatization factor of 1°C per 30 years (Dessai 2002) was included to allow for physiological and behavioral adjustment to higher temperatures. The same model was also applied for three CIRCE simulations (ENEA, MPI and IPSLglo) (Fig. 5.17). Even after adjustment for adaptation, there are significant increases in excess heat-related mortality for the periods 2021–2050 (Figs. 5.16 and 5.17) and

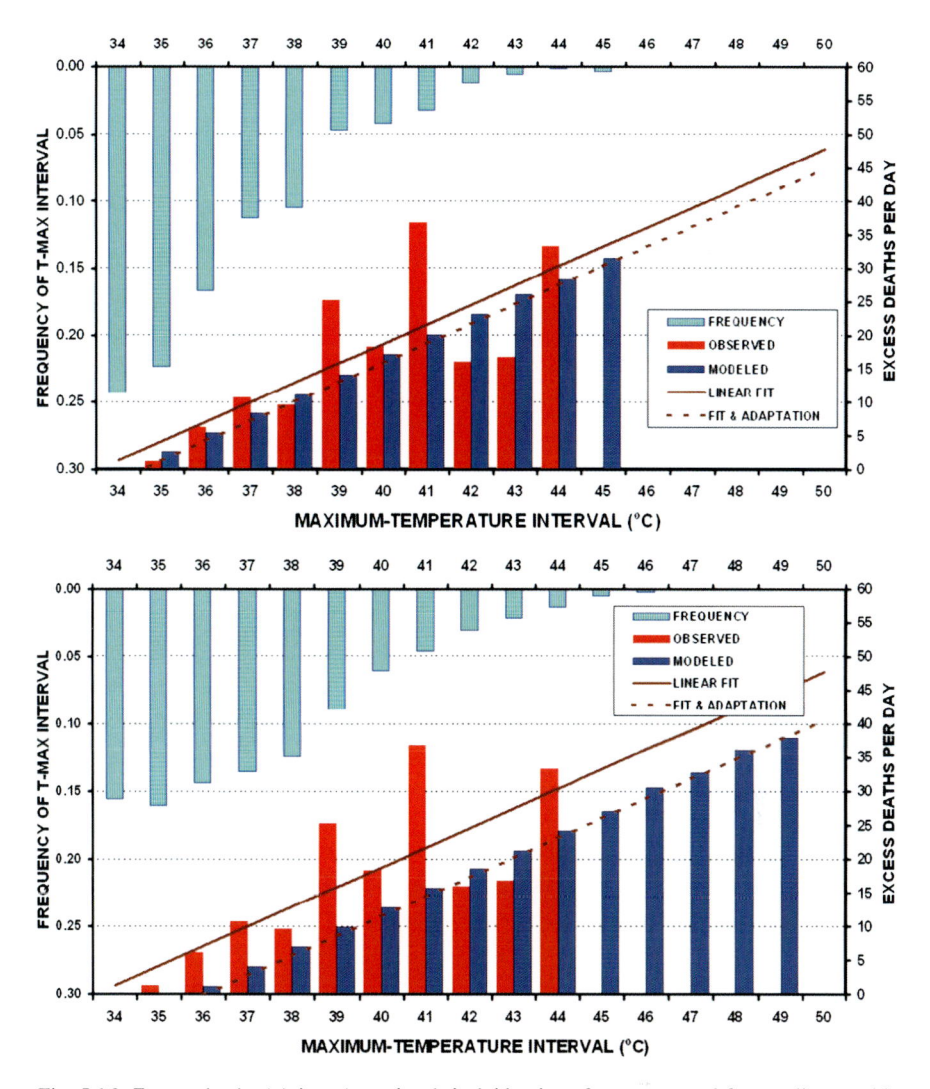

Fig. 5.16 Excess deaths (*right axis; red* and *dark blue bars* for present and future climate with adaptation, respectively) and daily temperature frequencies (*left axis; light blue bars*) in Athens, for the future periods 2021–2050 (*top*) and 2071–2100 (*bottom*), respectively, for the RACMO2 RCM ENSEMBLES simulation

especially for the period 2071–2100 (Fig. 5.16). This increase is evident in the figures by the shift to the right and the longer tails of the simulated color bars compared with the red observed bars. There are, however, substantial differences in mortality projections for 2021–2050 in the three CIRCE models (Fig. 5.17). In particular, the IPSLglo model fails to capture the high summer temperatures observed in Athens (see Sect. 5.4), and thus exhibits much lower values of excess mortality in the future climate. ENEA has a much better fit to observations, followed by the MPI model, providing greater confidence in the mortality projections.

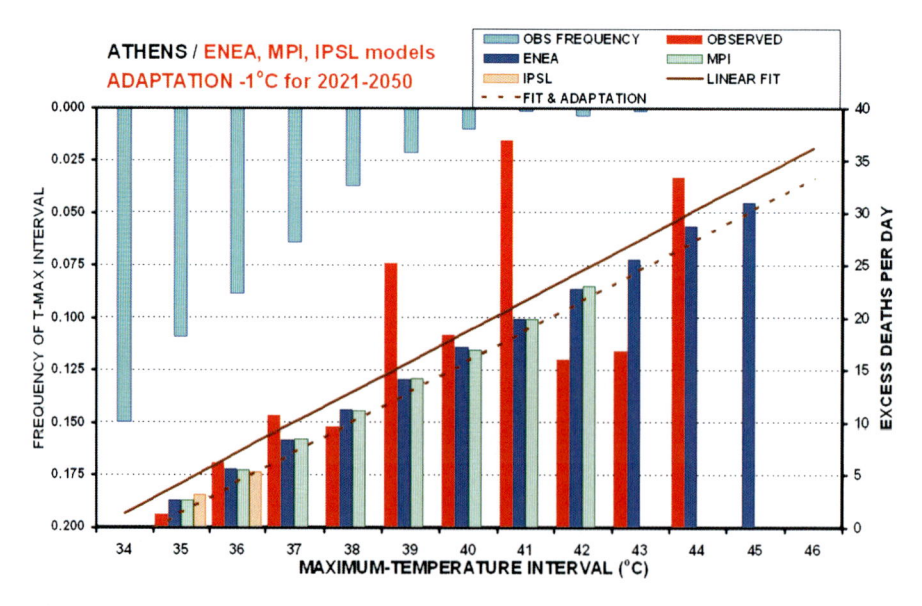

Fig. 5.17 Excess deaths (*right axis*; *red, dark royal blue, green,* and *light orange bars* for present and future climate with adaptation, respectively) and daily temperature frequencies (*left axis*; *light blue bars*) in Athens, for the future period 2021–2050 derived from three CIRCE multi-model simulations (ENEA, IPSLglo and MPI)

Energy Demand

Consumption of electricity is particularly sensitive to weather (in particular temperature), since large amounts of electricity cannot be stored and thus electricity that is generated must be instantly consumed. Average daily electricity demand in most European countries shows a single peak during winter months (Hekkenberg et al. 2009). In Europe, only Spain, Portugal, Italy and Greece currently show an additional peak during summer months (Hekkenberg et al. 2009; Giannakopoulos and Psiloglou 2006).

In order to explore potential future changes in energy demand, hourly energy consumption data were obtained from the Strategy and Planning Department of the Public Power Corporation of Greece for the Greater Athens area. These data are for total hourly residential and commercial electricity consumption (kWh) spanning the 9 years from January 1993 to December 2001. Over this period, energy consumption shows a clear upward trend that has largely been attributed to economic growth (as indicated by gross national product) and increasing use of air conditioning. However, at the same time, short-term fluctuations are non-linearly (negatively in winter, positively in summer) associated with the variation in daily mean temperature.

The variation in daily energy consumption with mean daily air temperature for 2001 is shown in Fig. 5.18. During December, the maximum values of energy consumption are related to the occurrence of the lowest temperatures. During the transition season of March/April, air temperatures are constantly rising but remain

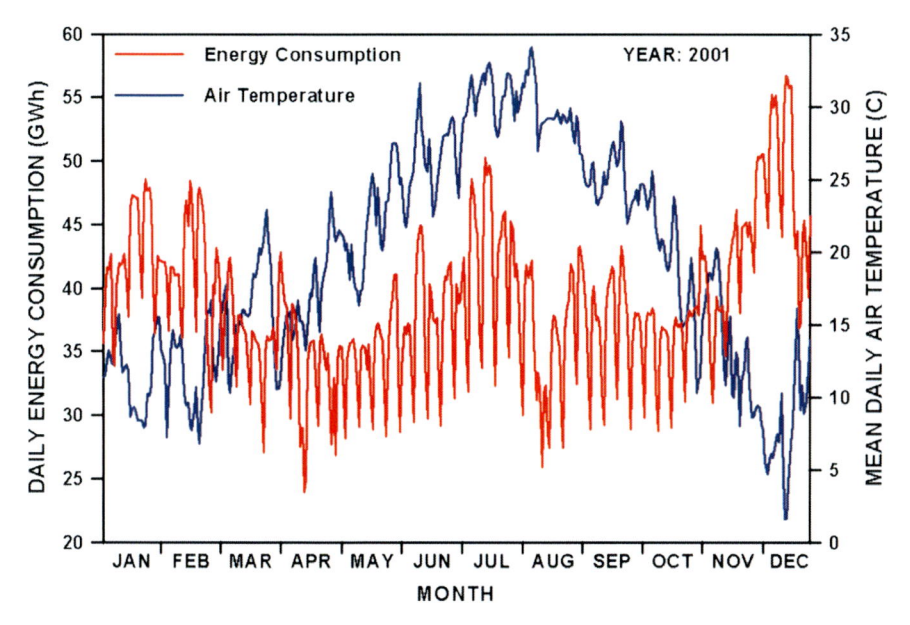

Fig. 5.18 Variation in Athens daily energy consumption (*left axis, red line*) with mean daily air temperature (*right axis, blue line*) for 2001

within comfort levels so energy consumption levels remain nearly constant until about May before rising to a summer peak. Energy demand and consumption are, however, greatly reduced during weekends and holidays compared to working days (Fig. 4.2). The lowest values occur during the long Easter weekend ('Good Friday' to 'Easter Monday') and other fixed holidays (15 August, Christmas), irrespective of daily mean air temperature. Nonetheless, in general in Athens, energy consumption (for cooling) increases with increasing temperature above a threshold temperature of about 22°C (Fig. 4.3; Table 4.1), and below this value consumption (for heating) increases with decreasing temperature.

Future energy demand in Athens was estimated using temperature output from three CIRCE models (ENEA, MPI and IPSLglo), and the RACMO2 RCM ENSEMBLES simulation for the A1B emissions scenario and a simple extrapolation of the non-linear trend between temperature and energy load for the present day. It is assumed that technology use (UNDP 2002) is the same in the control (1961–1990) and future periods (2021–2050 and 2071–2100). Figure 5.19 shows that for the colder half of the year (November to April), energy demand decreases. The gain in energy demand is about 2% for 2021–2050 and about 5% for 2071–2100 (RACMO2 results, not shown). For the warmer half of the year (May to October), an increasing trend is evident: a 5% increase in demand for 2021–2050 and 15% for 2071–2100 (RACMO2 results, not shown). The increases are double these values in the hot summer months of July and August when the demands for air conditioning are at their peak. For the CIRCE simulations, the summer gain in demand is similar (perhaps slightly less) in percentage terms to the winter decrease (Fig. 5.19), whereas it is greater for the RACMO2 simulation (not shown).

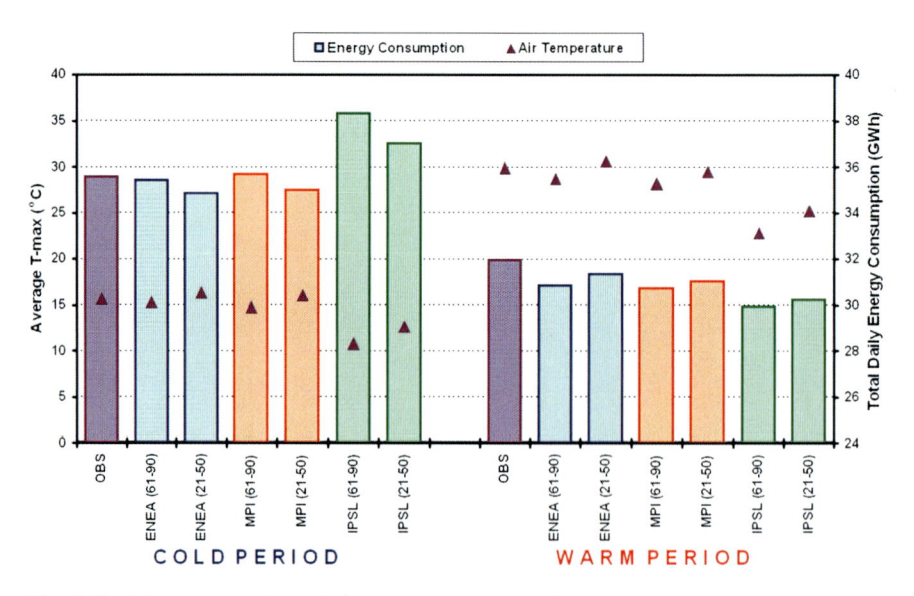

Fig. 5.19 Athens energy demand (*bars*, *right axis*) and daily maximum temperature (*triangles*, *left axis*) for the colder (*left*) and warmer (*right*) halves of the year for 1961–1990, 2021–2050 periods for CIRCE simulations (ENEA, MPI, IPSLglo)

5.3.3.2 Gulf of Gabès Coastal Case Study

An assessment of CIRCE climate projections (Sect. 5.2) for the Gulf of Gabès shows clear warming in both the mean and extremes of air temperature (Figs. 5.1, 5.2 and 5.3). Associated with this atmospheric warming, sea surface temperatures in the Gulf are also projected to warm significantly at about 0.2°C per decade for the period 2021–2050 relative to the period 1961–1990 and 0.4°C per decade by the middle of the twenty-first century (Fig. 5.10). There is some evidence of a non-significant increase in salinity out to the middle of the present century, with a decrease towards the end of the century (not shown). The CIRCE models show an increase in sea level of around 10 cm for the period 2021–2050 relative to the period 1961–1990 in the Mediterranean basin due to the steric effect (see more detailed discussion in Sect. 5.2). Mean annual precipitation (Fig. 5.8) is projected to decrease in the Gulf of Gabès at a relatively low rate compared to some of the other case studies (about − 4% per decade in the period 2021–2050), while wind speed and relative humidity (not shown) show no significant change. These atmospheric and marine changes are expected to lead to a range of biogeophysical and socioeconomic impacts in the Gulf of Gabès. Below, potential impacts for marine ecosystems and the tourist industry are presented using selected impacts and vulnerability indicators.

For marine ecosystems, growth in the number of alien (invasive) species was estimated by fitting an exponential relationship to the observations (adjusted to remove a bias introduced by the increased research, and hence reporting activity,

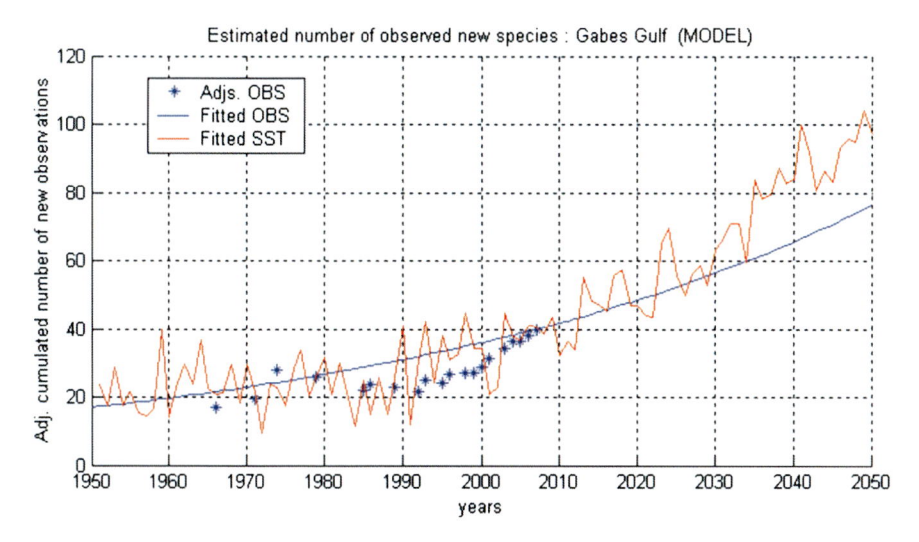

Fig. 5.20 Estimated number of observed alien species in the Gulf of Gabès (*red line*) obtained as a fitted exponential function to SST (Fitted SST = 42.8*SST-741.6). An exponential fit to observations (*blue line*) with a trend of 15%/decade is also shown. The observations (*blue symbols*) are the cumulative number of new species recorded, (see Harzallah et al. 2010) and are adjusted to correct for the increasing number of studies from a reference year, 2007

in the area). The underlying observations are shown in Fig. 3.7, while the estimated future growth as a function of the projected ensemble-mean SST warming (Fig. 5.10) is shown in Fig. 5.20. The projected increase in the number of alien species at a rate of +43% per 1°C warming results in an estimate of nearly 106 new species by 2050 compared with the 2007 baseline. Such an increase would have further 'knock on' effects on the region's marine biodiversity and the economically important fishing industry.

Tourism is of growing importance to the regional economy in the Gulf of Gabès, particularly for locations such as the island of Djerba. A Tourism Seasonal Shift Index based on a threshold mean temperature of 23°C defining the start and end of the tourist season was calculated using output from five CIRCE models for the period 1950–2050. The simulated start and end dates (Fig. 5.21) suggest a further lengthening of the summer tourist season (for the period 2021–2050: 3 days earlier start, 13 days later end, 16 days longer duration). Such a change could bring benefits to tourism activity in the southern gulf, particularly in the autumn season.

The future climate 'attractiveness' for tourism in the Gulf of Gabès was also assessed using a Daily Climate Tourism Potential Index (DCTPI; Henia and Alouane 2007; see also Sect. 4.4.2), for which five classes are defined: 'highly unfavorable', 'unfavorable', 'acceptable', 'favorable', and 'highly favorable'. Model data were adjusted to those of the observation period 1973–2003 for Djerba (Fig. 4.18) to help correct for climate model biases (see Sect. 5.4). The results (Fig. 5.22) suggest that

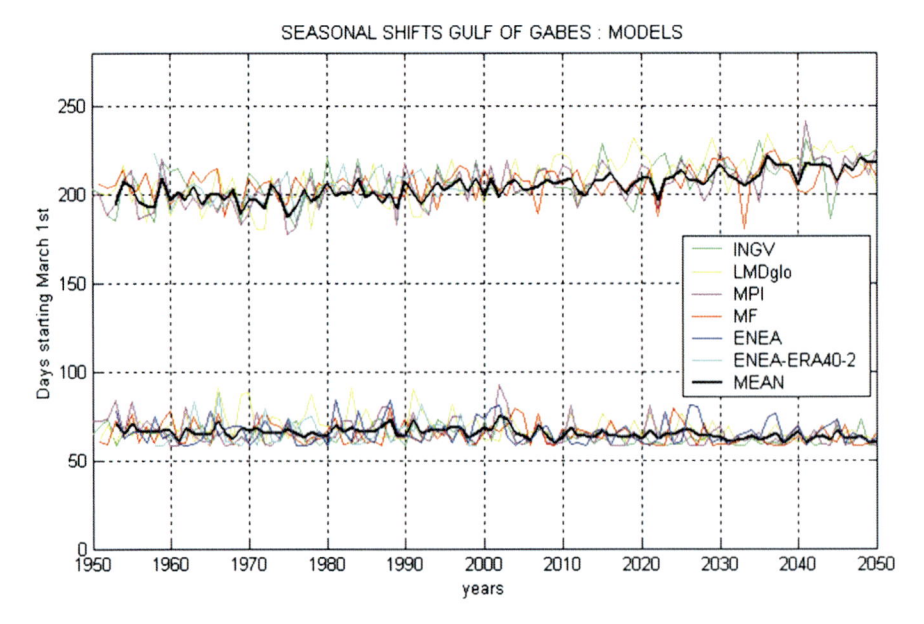

Fig. 5.21 Tourism Seasonal Shift Index (T > 23°C): the start and end of the summer season calculated using output from a range of CIRCE models for the period 1950–2050. The ensemble-mean (based on all models except ENEA-ERA40-2) is also shown. Days are counted from 1st March. Prior to calculation, the mean daily temperature series from each model was adjusted to the mean observed temperature at Djerba during the period 1973–2003

the number of 'favorable' days will decline steadily, while 'acceptable' days increase. By the middle of the twenty-first century, nearly 22 days a year (largely in summer) will be downgraded, with respect to the present climate, from 'favorable' to 'acceptable'. In contrast, tourism will benefit from a small increase in the number of 'highly favorable' days in winter.

The socio-economic consequences of climate conditions becoming 'acceptable' rather than 'favorable' for tourism were then calculated theoretically on the basis of a devaluation of −20 Tunisian dinar per night (MEDD/PNUD 2008). All capital gains and losses are based on MEDD/PNUD (2008) and are reported relative to GDP (2007 levels), while employment loss is considered to be dependent on hotel capacity. The estimated economic impact for the region by the year 2050 relative to 1960 is a loss of capital and annual gains of 0.023 and 0.11% respectively (Fig. 5.23). Consequently, by 2050 there could be a direct and indirect reduction in employment of around 5 and 14,000 people, respectively. It should be noted that the socio-economic impacts are estimated only according to the hypothesized effects of temperature warming on tourist attractiveness, other system drivers and pressures (e.g., a longer tourist season, beach erosion and loss, availability of freshwater, and the cost of air travel) are not considered.

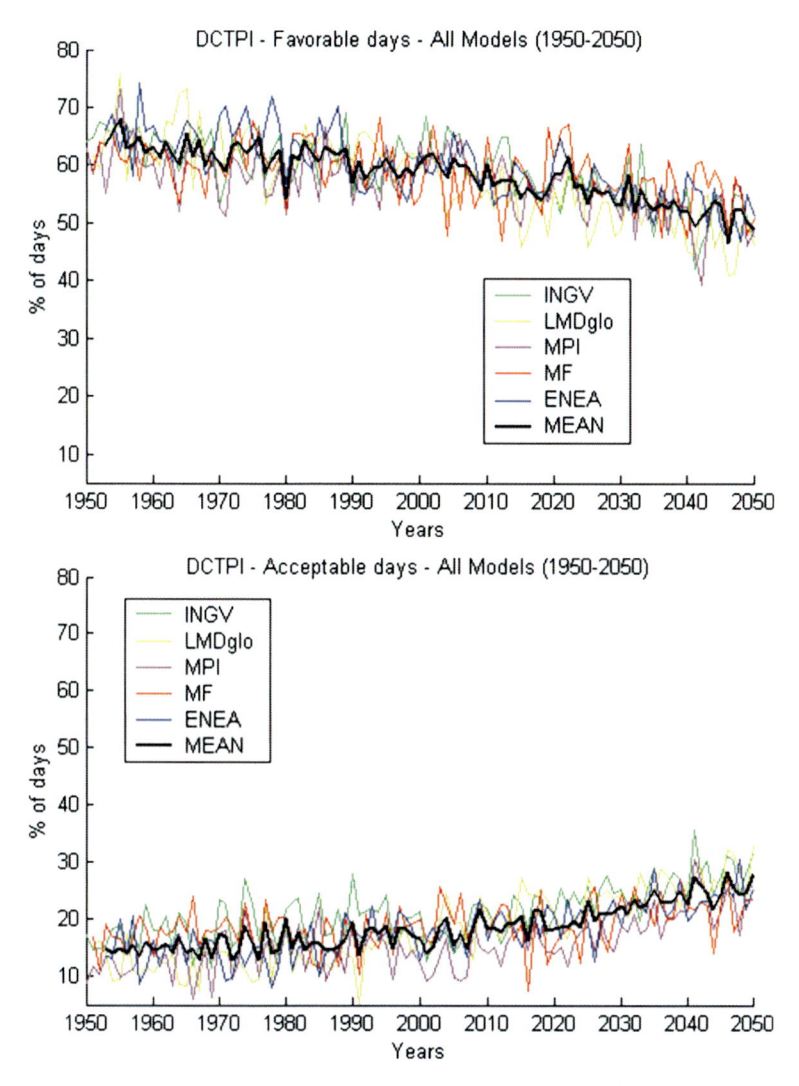

Fig. 5.22 DCTPI (Daily Climate Tourism Potential Index) time series (1950–2050) for the classes 'favorable days' (*top*) and 'acceptable days' (*bottom*) for the Gulf of Gabès. The shown series are calculated from the output of five CIRCE coupled models, and the ensemble mean

5.3.4 Summary and Discussion of Key Projected Changes

The CIRCE integrated case-study assessments aimed at using diverse types and sources of knowledge (qualitative and quantitative) to understand climate hazards, impacts and vulnerability at a local/regional level (see Figs. 4.2, 4.6 and 4.11). In this section, some of the key projected changes of relevance to the integrated case studies are discussed first for biogeophysical systems, and second for social systems

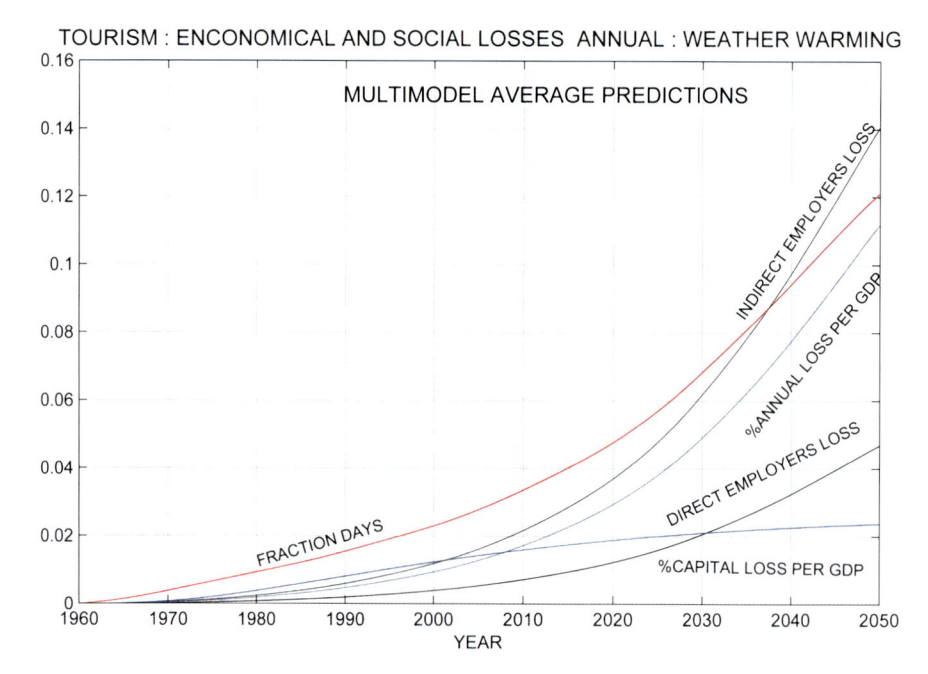

Fig. 5.23 A theoretical simulation of socio-economic impacts of days becoming 'acceptable' in the Djerba-Zarzis tourist area for the period 1960–2050, based on the ensemble average of five CIRCE coupled models for the Gulf of Gabès area (the average of the different model variables used to calculate the percent days becoming 'acceptable' are adjusted to that of the observations for the period 1973–2006). Results are shown as annual percentages of 2007 GDP. Projections of the direct and indirect employee loss should be multiplied by 100,000. Economic losses are based on 2007 rates

and communities in the Mediterranean. Finally, the case studies conceptual framework (the CCIF) is presented as a useful tool for integrating different types of information across sectors, for highlighting knowledge gaps, and for identifying strategies for adaptation.

5.3.4.1 Future Changes in Biogeophysical Systems

While a general northward shift in evergreen forest and general increases in productivity of durum wheat are projected in CIRCE work (Part III), local impacts will depend on a complex interplay of soils, water resource management, and farm-level inputs and management – as illustrated in the Judean Foothills and Tel Hadya rural case studies. Land use is also changing across the Mediterranean region due to non-climatic drivers. In northern regions, rural depopulation and abandonment of agricultural land is a characteristic process. In contrast, areas in the south and east of the Mediterranean region (such as Syria, Algeria, Tunisia, and Egypt) are experiencing pressures of rapid population growth, and increasing use (and overuse) of rural land for livestock, cropping and firewood (Part III, Sect. 2.4). At a regional level, two rural case studies within the same country (Italy), typify these diverse processes: land abandonment in Tuscany and intensification of agriculture in the Apulia case study. All of the rural case studies have considered the potential impacts of climate change for rain-fed wheat crops. In each case, the key climate hazards identified are heat stress, drought and a reduction in water availability. However, the level of vulnerability to climate change depends not just on the future extent of the climate hazard, but on the complex mix of environmental and social vulnerability and stressors. For example, local differences in farm management, water resource type (ground water, surface water, or reclaimed wastewater) competition for freshwater resources, slope and geology, and land use; and in the capacity to adapt (e.g., differences in financial resources, policy, and technological innovation – see Chap. 6).

Ecosystem vulnerability is also spatially diverse across Mediterranean regions. Semi-arid ecosystems are particularly vulnerable to climate variability and change, with high rates of population growth adding increasing pressure on natural resources and increasing the likelihood of overexploitation of provisioning ecosystem services potentially resulting in desertification and low land productivity. In areas lacking effective policies or investment in environmental protection and management, ecosystem vulnerability is further heightened. Nature reserves in the Gulf of Oran are ecologically important littoral sites and buffer zones (Sect. 4.4.3), but demonstrate substantial fragility in the face of climate change and marine pollution. Atmospheric and marine changes may already have contributed to the appearance and persistence of invasive marine species and a reduction in prey species distribution and abundance has potential damaging consequences for predatory species such as dolphins, and for the fishing industry. Terrestrial ecosystem vulnerability to climate change in the Mediterranean is highlighted in the Judean Foothills rural case study (Sect. 4.3.1). Ecosystem changes (presence or absence of vegetation types) along a rainfall gradient show the potential vegetative impacts of a change from a more moderate precipitation regime (tall woody vegetation and forest) to a semi-arid regime (sparse dwarf shrubland).

5.3.4.2 Future Changes in Social Systems and Communities

Climate projections for the Mediterranean region (Sect. 5.2) suggest future threats to urban areas, most notably and robustly in the form of increased risks of high-temperature extremes such as heatwaves, and will inevitably aggravate other climate-related impacts, such as water resource availability, saltwater intrusion, air pollution and peri-urban forest fire risk. Socio-economic sectors will also be impacted, for example, human health risks are expected to increase with more frequent and extreme high temperatures, droughts, and air pollution events, together with associated challenges of rising energy demand for cooling and shifts in the seasonal pattern of tourism.

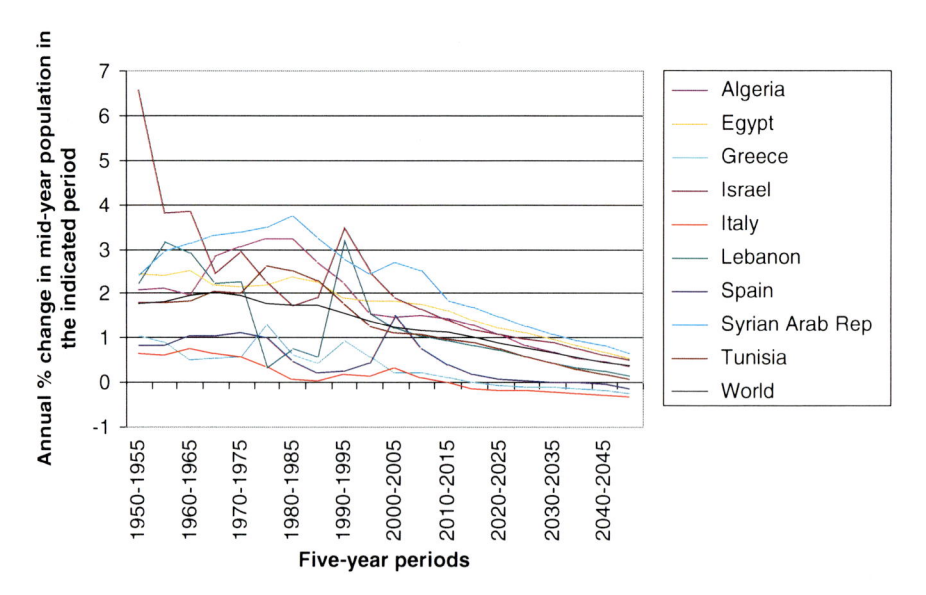

Fig. 5.24 Total population growth rates for the case-study countries (Source: Population Division of the Department of Economic and Social Affairs of the United Nations Secretariat. 2007. World Population Prospects: The 2006 Revision. New York: United Nations)

It is important to consider non-climate drivers of change in biogeophysical and social systems, such as population and economic change, alongside climate changes. While Fig. 5.24 indicates a future decline in population growth rates over the course of the next 50 years, projections for countries such as Greece, Israel and Tunisia, show continuing growth in urban population and declining rural population. In the short-term population growth is likely to exert additional pressure on land and resources particularly in the southern and eastern countries of the Mediterranean. Note that a number of different population scenarios and projections are available and have been used across the CIRCE project. The energy-related work reported in Part IV, Chap. 14, for example, uses the International Labor Organisation (ILO 2009) as a source for population projections for North African countries. These are higher than the UN projections shown in Fig. 5.24 by about 1% per annum.

Critical water issues in the case-study locations are related to increasing demand in regions of population growth (e.g., in Algeria, Egypt, Israel, Lebanon, Syria and Tunisia), and increasing demand in areas of more intense agricultural use (e.g., Apulia). Access to adequate water resources is a basic need and is pivotal to all sectors of society, from ecosystem services, to industry and services, human health and well being. In the conceptual framework for individual case-studies, freshwater resources is a crucial integrating indicator for several regions (Beirut, Alexandria, Tel Hadya, Judean Foothills, Apulia, and Tuscany), and water scarcity or inequalities in access has economic and welfare implications. Since water integrates sectors and regions, promoting policies and activities which foster equality in water provision and greater efficiencies in water management are essential components to future sustainable development in the Mediterranean. Reclamation of wastewater from domestic and industrial sources, as well as floodwater storage are potentially large sources of irrigation water for agriculture and other water consumers. However, use of recycled wastewater requires building appropriate infrastructure and in countries such as Lebanon there are social and religious constraints on certain uses. Large amounts of water can also be saved through utilization of efficient irrigation technology in agriculture, such as dripping irrigation.

Within the countries of the Mediterranean, vulnerability to climate-related disease, injury and death vary spatially due to regional and local differences in current and projected climate hazards, and in health-care services, and in preparedness for extreme events such as heatwaves. An assessment of impacts and vulnerability in the urban case studies (Athens, Alexandria and Beirut) has highlighted the diversity in climate-related health hazards and social vulnerability between cities of the Mediterranean. In Athens, heat waves have readily discernible and direct health impacts because they result in a large number of deaths and affect relatively large, heterogeneous areas simultaneously. However, health outcomes vary according to the intensity, duration and timing of the heatwave event, and according to the characteristics of the population, disproportionately affecting the elderly, young people with pre-existing illness, and low-income groups. The threshold at which heat deaths occur also varies across the Mediterranean and is higher in cities of the southern and eastern Mediterranean where communities are more accustomed to warmer temperatures (Part IV, Chap. 13).

Synergistic effects of high ozone and heat extremes can increase the mortality rate in some cities. The Athens (see Sect. 5.3.3.1) and Alexandria urban case studies have both highlighted the increasing risk of ozone pollution and corresponding health concerns. High pollutant and aerosol loadings are of particular concern in Alexandria (Fig. 3.6), and in the wider West Nile area (although no future projections are available for this region from CIRCE air quality work). Peak pollution days typically occur during the spring and summer dust seasons when pollutants are transported from Europe. An estimated 10–25,000 people a year currently die in Cairo due to air pollution-related disease. Yet despite the health concerns, unregulated human activity and emissions persist, and the health risks of climate change are under-researched in the region. In a future warmer world, faster chemical reactions, increased biogenic emissions, and atmospheric stagnation could increase the risk of pollution episodes in Mediterranean cities (as suggested by the statistical modeling work on ozone pollution for Athens – see Sect. 5.3.3.1, Fig. 5.15).

Heat stress and water-borne disease are two key issues of concern in relation to human health in Beirut (Sect. 4.2.1). Projected changes in mortality due to heat-related stress and food- and water-borne diseases have been explored using CIRCE model and CMIP3 global climate model outputs. The results (not shown) indicate increasing trends in food- and water-borne disease Changes in heat-related mortality have been explored using the 27.5°C threshold identified by El-Zein et al. 2004. A particular aim is to quantify the balance between increased warm-season and decreased cold-season deaths, and between mortality changes related directly to warmer temperatures and the projected increases in mortality from food- and water-borne disease.

The tourism industry makes an increasing contribution to the economy of Mediterranean countries, and particularly to coastal regions. *The United Nations Environment – Mediterranean Action Plan* (2005) has projected that tourism in the Mediterranean will rise from 246 million tourists in the year 2005 to around 637 million tourists by the year 2025. However, these estimates do not incorporate the potential impacts of climate change. Future inflows to tourist destinations in the CIRCE case-study regions are likely to be affected (directly and indirectly) by changes in average climate (in both the source and destination regions), changes in the frequency and intensity of extreme climate events (such as floods, droughts, and heatwaves), and by international and national climate change mitigation policies. Highlighted climate-related risks for tourism in the case studies include: beach erosion and loss (e.g., for the low-lying islands of the Gulf of Gabès), heat-related extremes (Athens, the Judean Foothills, Apulia, the Gulf of Gabès), atmospheric pollution and wildfires (Athens, Alexandria), increased competition for scarce water resources and saltwater intrusion of coastal aquifers (Gulf of Valencia, Gulf of Gabès, Alexandria, Beirut, Apulia), and loss of traditional landscape in Tuscany. However, the direction of response to projected climate change for Mediterranean resorts may not be all negative: the quantitative impact assessment of climate change in the Gulf of Gabès coastal case study (see Sect. 5.3.3.2) suggests that the projected warmer winters and simulated extension of a temperature-determined tourism season present opportunities in winter and in the transitional seasons, particularly in autumn. In the rural Apulia case study, a projected longer summer tourism season by the mid twenty-first century (2.2 days/decade) has the potential to offer new opportunities to regional tourism.

5.3.4.3 Linking Vulnerabilities and Impacts

The CIRCE case-study assessments of impacts and vulnerability show that climate variability and change have direct consequences on a variety of physical and social systems that in turn affect other physical and social systems through a cascading set of processes (Chap. 4). The development of an individual case study conceptual framework (the CCIF; see Sect. 1.3) attempts to organize these linkages and highlight important research issues regarding key climate hazards, vulnerabilities and impacts at a regional-to-local level. This approach also facilitates the identification and implementation of local and regional adaptation strategies (see Chap. 6).

Figure 5.25 shows the CCIF indicator linkages for the Gulf of Gabès case study for which a number of key hazard, impact and vulnerability indicators have been quantified for both the present-day and the future (Sect. 5.3.3.2). The linkages relating to the tourism sector have been studied in some detail. This sector may be negatively impacted through beach erosion, saline intrusion, degradation of marine ecosystem services, and the projected increase in heat extremes. However, climate change may present some opportunities for regional tourism due to the projected increase in days 'favorable' to tourism in winter, and an extension of the tourist season into the autumn. Another set of linkages that has been explored in detail concerns alien

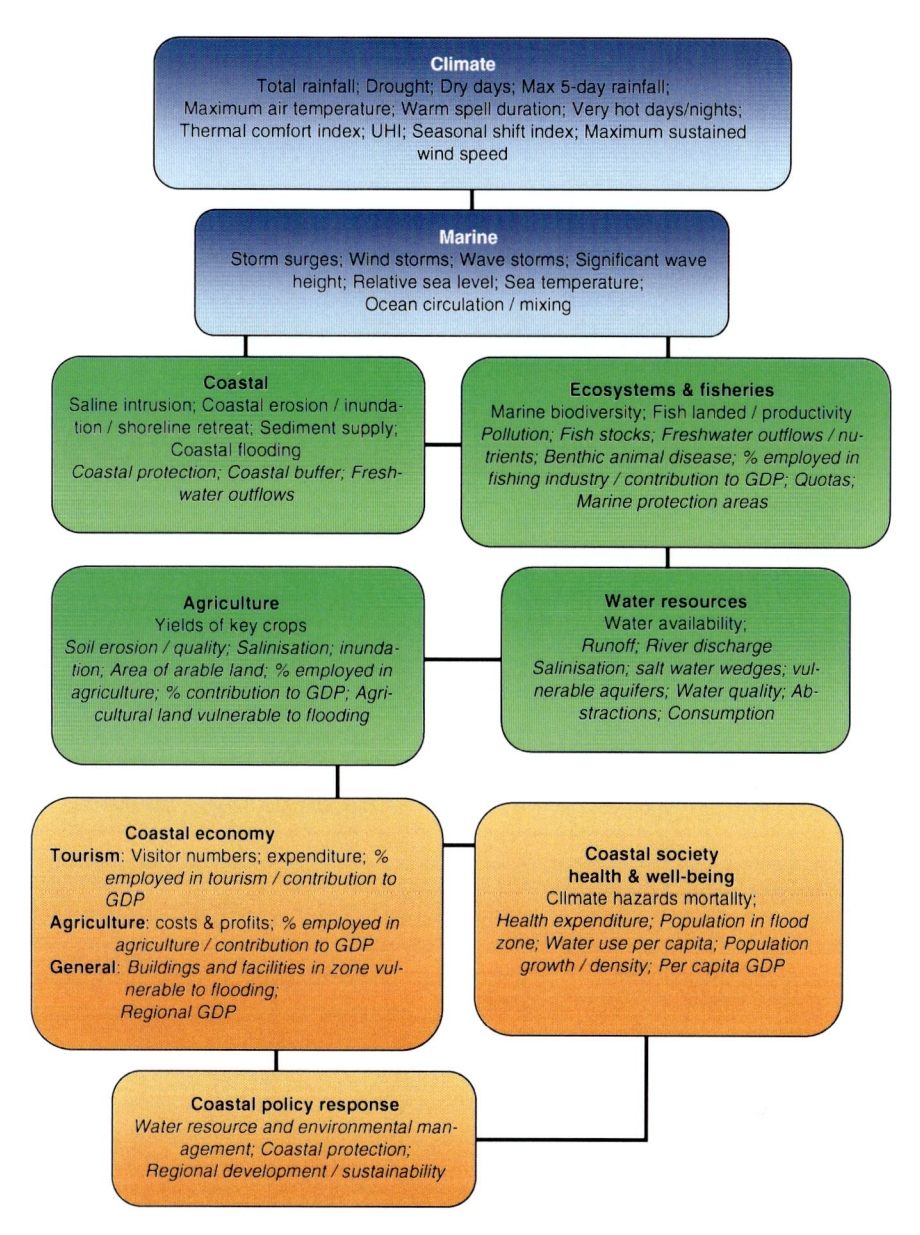

Fig. 5.25 Indicator linkages for the Gulf of Gabès, Tunisia. The *blue boxes* denote key climate and marine drivers, the *green boxes* denote key biogeophysical impacts and vulnerabilities (*italics*); the *orange boxes* denote key social impacts and vulnerabilities (*italics*)

species and fishing activity vulnerability (Fig. 5.26). It is surmised that an increase in the annual maximum swell height (Fig. 4.14) together with sea-level rise and human-derived shoreline modification have accentuated the vulnerability of the coastline to climate change. In parallel, marine ecosystems in the region are particularly vulnerable

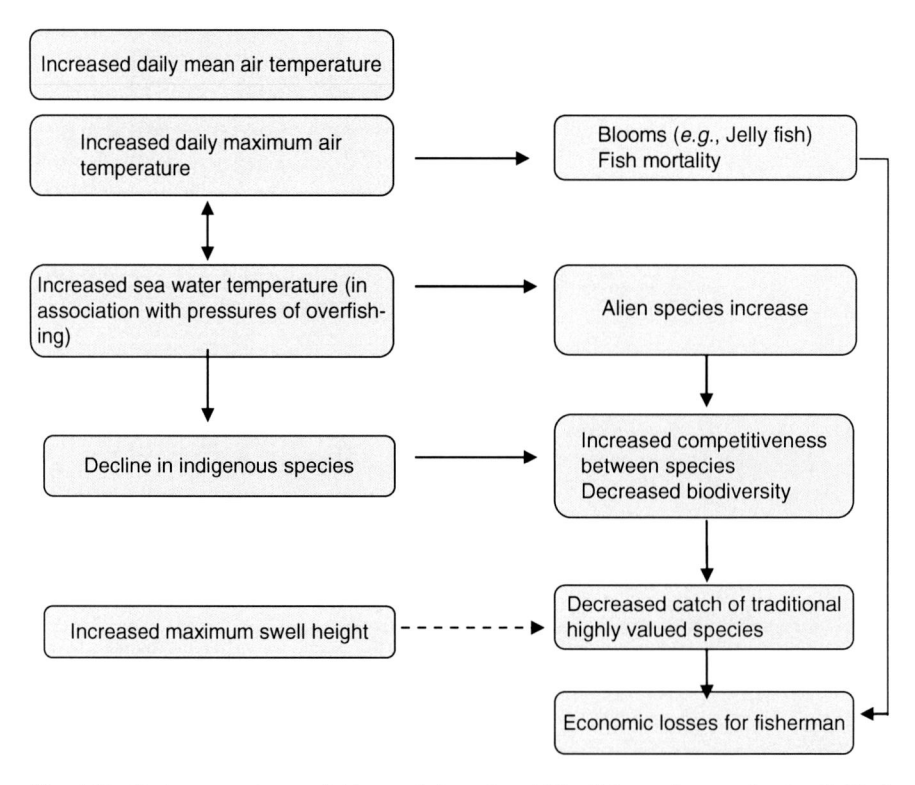

Fig. 5.26 Marine ecosystem – fishing activity vulnerability linkage diagram for the Gulf of Gabès

to sea-surface warming (Fig. 5.10) due to the shallowness of the Gulf and as a consequence, and in association with social pressures of overfishing (declining fish stock favors the establishment of alien species), the number of alien marine species found in the Gulf is projected to increase in the future (Fig. 5.20). In turn, this is likely to damage the biodiversity of the Gulf through food-web alterations which in turn have potential negative consequences for the economically important fishing activity in the region.

Within the financial and time constraints of the CIRCE project, it has not been possible to fully explore all the linkages identified in Fig. 5.25 and only a few of the indicators have been quantified for both the present day and the future (Table 5.3). Others have been considered using more qualitative and inferential approaches. It may be possible to fill some of the quantitative gaps in work beyond the end of the CIRCE project. Nonetheless, the linkage diagrams constructed for the Gulf of Gabès (Fig. 5.25) and the other case studies (Fig. 3.4, for example, shows the linkage diagram for the Judean Foothills), provide a valuable framework for discussions with regional stakeholders and decision makers (Sect. 6.3). Where quantitative indicators have been produced for the future, these are inevitably subject to a number of uncertainties which are discussed in the next section.

5.4 Assessment Uncertainties

The climate projections which underlie the case-study impact and vulnerability assessments are inevitably subject to uncertainties (Part I, Sect. 3.5). In terms of the climate model response to prescribed greenhouse gas and natural forcings, the CIRCE modeling team identify three types of uncertainty: unpredictability, structural uncertainty and value uncertainty (Part I, Table 3.12). These uncertainties can be addressed by taking either a multi-model or a perturbed-physics ensemble approach. The former approach samples inter-model or structural uncertainty and is the one adopted in CIRCE, while the latter approach samples intra-model or parameter uncertainty. The CIRCE multi-model ensemble is, however, rather small – six models in total (Table 5.1). This is because it was considered important to focus on atmosphere–ocean coupling over the Mediterranean, something not previously incorporated in simulations for the region. Thus a limited number of coupled regional simulations were undertaken within the project, with some additional analysis of the larger ENSEMBLES un-coupled RCM ensemble.

Before using any climate model output for the construction of future projections it is important to consider its reliability. A 'necessary but not sufficient' guide is provided by the ability of models to reproduce observed climate. Thus a number of evaluation or validation studies were undertaken as part of the CIRCE modeling work. In the case of the land-surface climate, a number of systematic errors were identified, with the CIRCE ensemble generally colder than observations in both winter and summer by about 2°C (Part I, Sect. 3.2.1). In terms of systematic biases in precipitation, the CIRCE models simulate an excess of precipitation over central Europe in both seasons and a lack of precipitation in the Alpine region, Middle East and, in summer, in the area of the Black Sea.

These biases are evident in comparisons of observed and simulated (ENEA and IPSLglo models only) maximum and minimum annual temperature (Fig. 5.27) and annual precipitation (Fig. 5.28) undertaken for the urban case studies. The cold bias is most clearly evident in both models in the case of minimum temperature for Athens and Beirut. In some cases, however, temperatures are well simulated (e.g., maximum temperature for Athens in the ENEA model) or are overestimated (e.g., maximum temperature for Alexandria in the ENEA model). Precipitation for Alexandria is clearly underestimated but appears better simulated for Athens. The models are not, however, able to simulate extreme years: there are no simulated values comparable to the exceptionally high total annual precipitation observed in Athens in 2006. In the case of Beirut, the observed inter-annual variability of precipitation is very large and there are concerns about the reliability of the observations, which makes it difficult to draw conclusions about model performance in this case.

In order to minimize the effects of such biases, the CIRCE projections are generally presented in the case-study work as the difference between the future (2021–2050) and control (1961–1990) periods. In the case of extreme events (Figs. 5.3, 5.4, 5.6, 5.7 and 5.9), biases in the absolute magnitude of events (but not in the shape of the

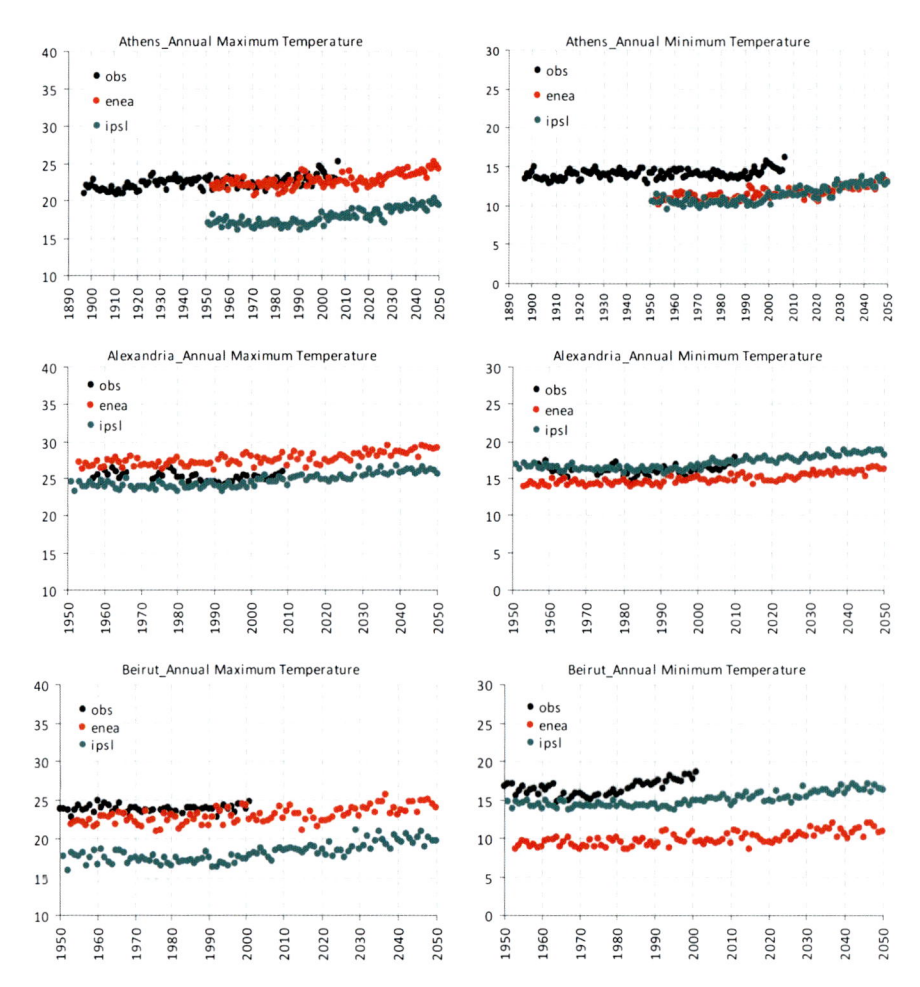

Fig. 5.27 Time series of mean annual maximum (*left*) and minimum (*right*) temperature (°C) for the urban case studies from observations (*black*) and the ENEA (*red*) and IPSLglo (*green*) CIRCE models. The models run from 1950 to 2050

distribution) are minimized by using percentile thresholds calculated from the model control period rather than from observations. Fixed, absolute thresholds are, however, rather easier to understand than percentiles. Thus, the Athens impacts studies (Sect. 5.3.3.1) generally used absolute temperature thresholds to give a clearer picture of the underlying impact to the relevant stakeholders. As an example, the IPSLglo model has a particularly large cold bias (Fig. 5.27) and projects virtually no increase in mortality when health impacts are considered (Fig. 5.17).

Some impact models, such as crop growth models, require daily time series as input data and simulated processes can be extremely sensitive to the absolute magnitude of input variables, such as temperature, particularly at crucial stages of phenological development. Thus in preliminary CIRCE crop-modeling work, the

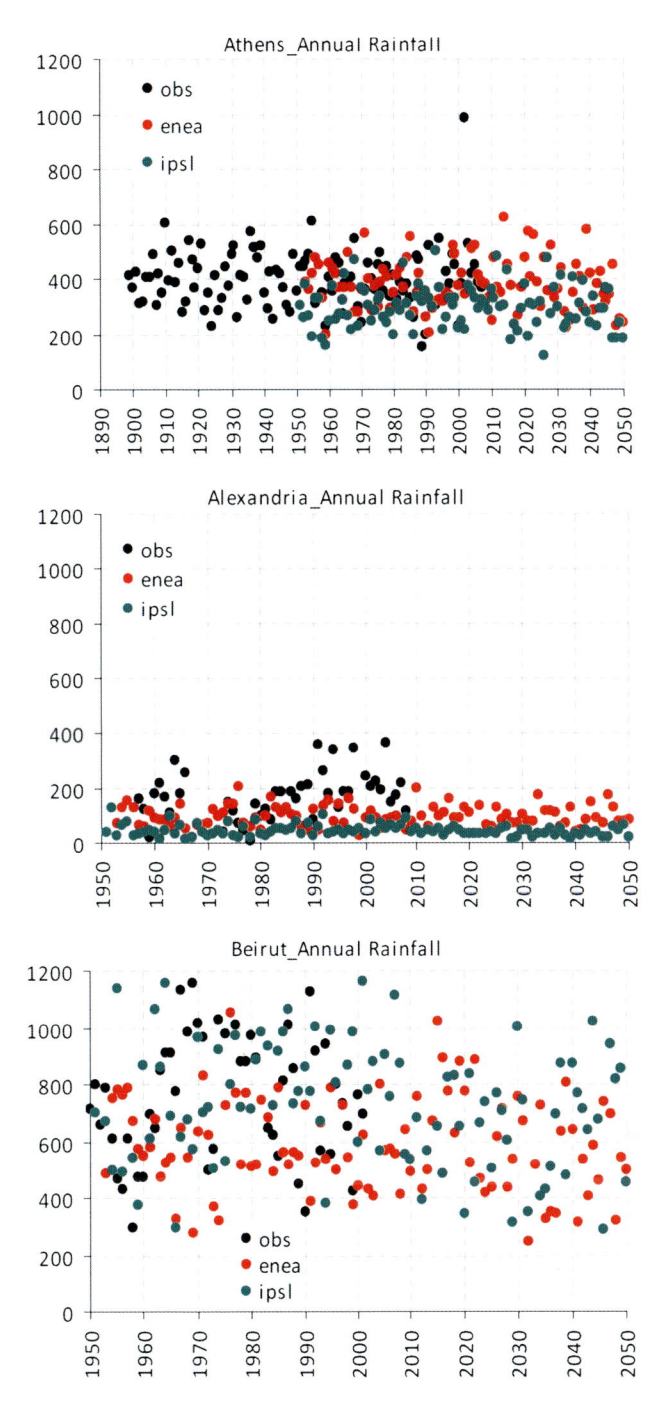

5 Integration of the Climate Impact Assessments with Future Projections

Fig. 5.28 Time series of total annual precipitation (mm) for the urban case studies from observations (*black*) and the ENEA (*red*) and IPSLglo (*green*) CIRCE models. The models run from 1950 to 2050

Tuscany group used the locally-calibrated LARS-WG weather generator (Semenov 2008) to stochastically generate homogeneous synthetic weather series for the present and future forced by GCM outputs. When forced by the HadCM3 GCM, the weather generator output indicates a general and progressive reduction of precipitation in Tuscany to the end of the century, particularly during spring and summer, together with a steady increase in temperature which is strongest in summer. The Tel Hadya group adjusted daily temperature, radiation and precipitation output from the ENEA model using cubic spline functions to fit the simulated to the observed data. Bias correction was also used in the Gulf of Gabès case study to adjust daily mean temperature output from the CIRCE models to observations for Djerba (Figs. 5.21 and 5.23). For the Gulf of Oran case study, a neural network technique was used to calibrate daily output from the ENEA model with observations. For all these bias correction methods, the adjustments or corrections are calculated using present-day data and then applied to model output for the future, on the assumption that the biases are unchanged in the future.

Biases are also evident in the sea-surface characteristics simulated by the CIRCE models (Part I, Sect. 3.3). A cold bias of about 2°C in SST is attributed either to air-sea fluxes being too cold during the spin-up coming from the atmosphere or to colder than observed waters being advected into the Mediterranean basin at the Strait of Gibraltar. Similarly, the models tend to underestimate SSS (by about −0.4 psu during the 1961–1990 control period) due to the advection of fresh surface waters from the Atlantic. Validation (and hence bias correction) of SSH is problematic due to temporal and spatial variability and lack of observed data.

Ideally, all six CIRCE climate model simulations would be used for the case-study work. However, constraints of time and resources meant that this was not possible (although work on processing model output is ongoing at the time of writing this volume). The analysis of mean temperature and precipitation and temperature extremes presented in Sect. 5.2 is based on five of the six-available models (Figs. 5.1, 5.2, 5.3, 5.4, 5.6, 5.7 and 5.8). When using these climate indicators it is important to consider the ensemble range as well as the ensemble mean (hence both are shown in the figures). The precipitation-based indicators for the urban case studies (Fig. 5.9) are based on only two of the CIRCE models: ENEA and IPSLglo – so caution is needed. Caution is also needed in using the biogeophysical and vulnerability indicators for Athens presented in Sect. 5.3.3.1. The fire-risk analysis is based on six RCM runs from the ENSEMBLES project and two CIRCE RCM runs, while the air pollution, human health and energy demand analyses are based on a single ENSEMBLES RCM and three CIRCE model runs. However, these analyses use projected changes in temperature which are generally more robust than precipitation changes. All the CIRCE models (and ENSEMBLES RCMs) agree in the direction of temperature change: an increase in both mean and high temperature extremes, although there is some uncertainty in terms of the magnitude of change (Sect. 5.2).

While the CIRCE models generally agree in projecting a decrease in mean annual precipitation across the case studies (Fig. 5.8), the uncertainties are greater at the seasonal scale (Part I, Chap. 3) and particularly with respect to extremes (see Fig. 5.9 and discussion in Sect. 5.2). This is due, in part, to the higher inter-annual variability of precipitation, but also to the higher spatial variability of precipitation and the finer spatial scale of the underlying physical processes – particularly with respect to convective events. In order to properly model convection and the diurnal cycle, non-hydrostatic mesoscale models must be run at resolutions in the order of 1–5 km in order to fully resolve cloud processes (Hohenegger et al. 2008).

From the *albeit* rather limited analysis of precipitation extremes undertaken for the case studies, it is difficult to make definitive statements about the likely direction of change in the frequency and magnitude of heavy precipitation events. Some previous studies do, however, suggest increases in the intensity of heavy precipitation in all seasons, except summer over parts of the region, but with a decrease in other parts such as the Iberian Peninsula (Goubanova and Li 2007; Giorgi and Lionello 2008; Giannakopoulos et al. 2009). These studies tend to focus on the end of the twenty-first century when the signal of change is greater compared with the 'noise' of natural variability than in the earlier (2021–2050) period considered in CIRCE. Support for an increase in the intensity of heavy precipitation, even where mean total precipitation decreases, is provided by RCM sensitivity runs undertaken for CIRCE (Part I, Sect. 3.4.3). These explore the influence of changes in temperature, specific humidity and large-scale circulation and suggest an increase in the intensity of heavy precipitation throughout the year, with the exception of the southwest Mediterranean region where intensity is reduced in late spring and summer. Note that these sensitivity experiments only provide information for aggregated regions and hence no spatial detail.

Changes in precipitation extremes and their potential implications for changes in flood risk were identified as relevant climate hazards for the urban case studies. Although some riverflow modeling was undertaken in CIRCE (Sect. 5.3.2) and coupled river-routing is implemented in some of the CIRCE climate models (Part I, Sect. 3.1.3), modeling of urban flood risk – which requires a mesoscale approach and detailed representation of the urban land surface and drainage systems – was not feasible within the project. Similarly, it was not possible to undertake mesoscale modeling of the urban heat island including detailed representation of the urban morphology. A number of RCM runs incorporating the urban surface and energy inputs in urban areas were, however, undertaken as part of the urban case study work in order to explore the sensitivity of the urban heat island to climate change (Box 4.1). Other mesoscale and microscale climatic influences which could not be explored include the differences in north- and south-facing slopes and the strong precipitation gradient in the Judean Foothills.

The large inter-annual variability of precipitation makes it difficult to identify statistically significant trends in observed (Chap. 4) or simulated (Fig. 5.28) precipitation-based climate indicators. For temperature, the observed trends and climate change signal are stronger. Figure 5.29 shows observed and simulated (ENEA and IPSLglo models) mean annual maximum and minimum temperature for the urban case studies – in this case, as anomalies from the 1961–1990 baseline in order to correct for the model biases (Fig. 5.27). In general, the projected changes appear as an extension of the observed positive trends. In some cases, most notably for Athens maximum temperature, the projected changes do not extend much beyond the upper

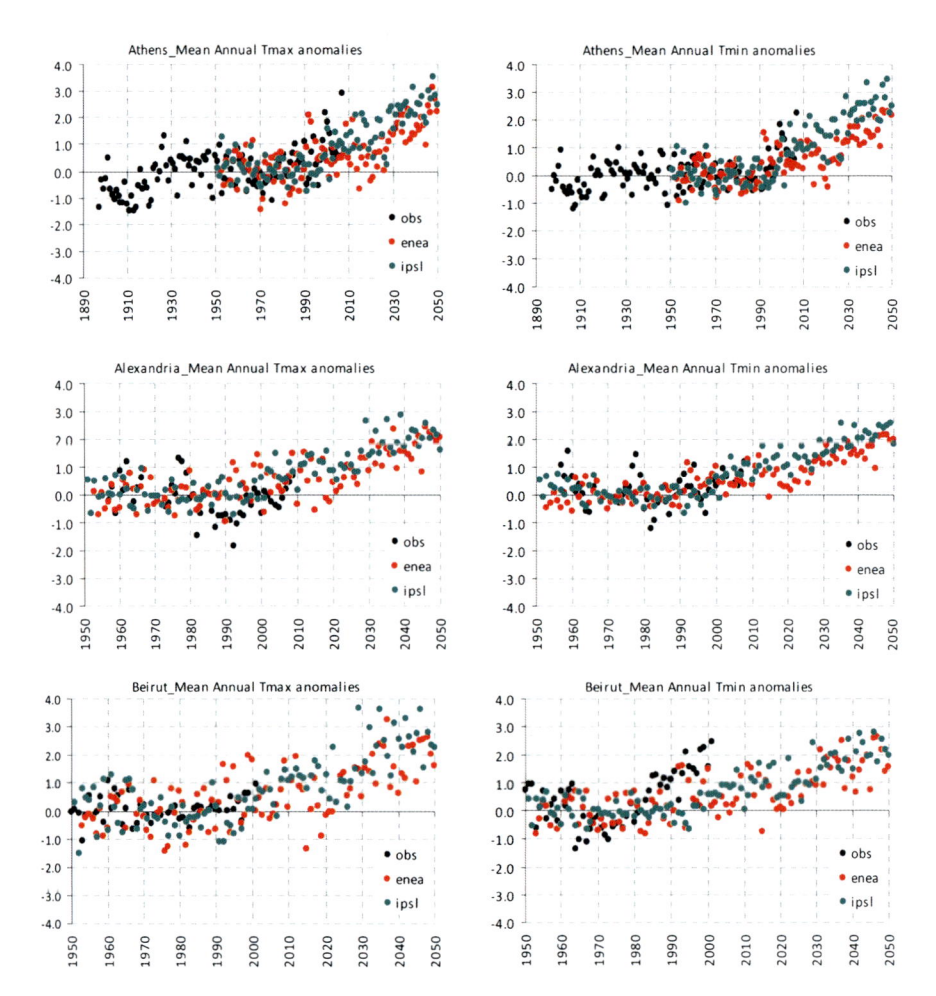

Fig. 5.29 Time series of anomalies (with respect to 1961–1990) in mean annual maximum (*left*) and minimum (*right*) temperature (°C) for the urban case studies from observations (*black*) and the ENEA (*red*) and IPSLglo (*green*) CIRCE models. The models run from 1950 to 2050

range of the observations. The influence of the hot summer of 2007 is evident in the 2007 observed Tmax value for Athens which appears as an outlier with respect to the observed series but falls centrally within the model range by 2050 (see also Box 5.1). Similar features are seen in the case of temperature extremes – very hot days and very hot nights (Fig. 5.30), although the inter-annual variability is greater and the upper range of the observations is not exceeded very frequently in the future. In the case of Beirut very hot nights, this is most likely a reflection of problems with the observed data which appears to be inhomogeneous after about 1980.

As part of the uncertainty assessment of the CIRCE climate models, mean seasonal responses were compared with those of the ENSEMBLES projections (Part I, Sect. 3.5.3). Probability distribution functions (pdf) were constructed from

CIRCE output for six Mediterranean cities including Athens, Tunis and Cairo which are the cities most relevant to the case studies. The pdfs of temperature change are generally centered on about +1 to +1.5°C and are somewhat skewed to the right, with tails extending out to about +5°C (Part I, Fig. 3.27). The pdf of precipitation change for Cairo is centered on zero, and those for Athens and Tunis just to the left of zero. All appear fairly Gaussian in shape. Care is needed in comparing the larger ENSEMBLES and smaller CIRCE ensembles, but it is concluded that introducing atmosphere–ocean coupling in CIRCE has increased the spread of the model responses (Part I, Sect. 3.6). It is also concluded that the CIRCE models have larger systematic errors compared with the ENSEMBLES RCMs. This is attributed to the fact that regional coupled models are a very recent numerical tool.

One source of uncertainty not addressed by either the CIRCE coupled model or ENSEMBLES uncoupled RCM ensembles is the choice of emissions scenario: both ensembles are based on SRES A1B. At least for mean temperature, the choice of emissions scenario is less important than model-related uncertainty until the second part of the twenty-first century (Hawkins and Sutton 2009) and thus is less of a concern with respect to the 2021–2050 scenario period used in CIRCE. While this conclusion with respect to the balance of uncertainty is certainly applicable to the choice of SRES scenario, it might not be applicable if aggressive mitigation scenarios were also considered. The ENSEMBLES project, for example, constructed the E1 mitigation scenario in which atmospheric concentrations of greenhouse gases are stabilized at 450 ppm CO_2-equivalent (van der Linden and Mitchell 2009). E1 was constructed using a reverse-engineered approach which has also been used to develop Representative Concentration Pathways (Moss et al. 2010) in advance of the IPCC Fifth Assessment Report. The CIRCE focus on a non-mitigation emissions scenario allows assessment of the impacts and damages which could be avoided through mitigation and also of the future climate that would need to be adapted to in the absence of mitigation.

The uncertainties in emissions and their conversion to concentrations of radiatively active species and radiative forcing lie 'upstream' of the climate model uncertainties in the overall cascade of uncertainty (Part I, Fig. 3.22). In the context of the CIRCE case-study integrated assessments, there are also uncertainties 'downstream' of climate modeling and the construction of climate change projections. Some of these are discussed below. In addition, there are uncertainties associated with all socio-economic projections used in the case studies – such as population projections (Sect. 5.3.4.2).

The quantitative impacts assessments undertaken for the CIRCE case studies, with the main exception of crop yield modeling in Tuscany and Syria, are generally based on statistical rather than process-based modeling. Such models are dependent on the reliability of the predictor (climate) data as well as that of the predictand (biogeophysical and socioeconomic) data. Most of the statistical models (such as those discussed in Sect. 5.3.3) are linear. Thus questions arise as to whether the linear statistical relationships identified for the present day can be extrapolated into the future (the assumption of stationarity). In part, this depends on the projected magnitude of change in predictor variable(s), so extrapolation

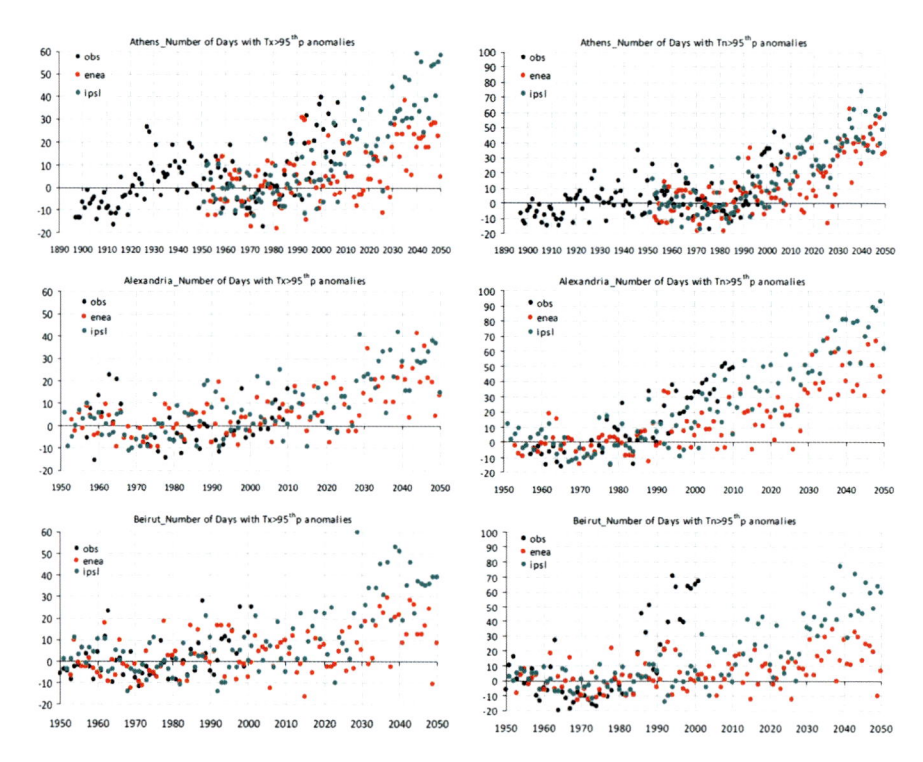

Fig. 5.30 Time series of anomalies (with respect to 1961–1990) in the number of very hot days (*left*) and very hot nights (*right*) for the urban case studies from observations (*black*) and the ENEA (*red*) and IPSLglo (*green*) CIRCE models. The models run from 1950 to 2050

may be less of an issue for the near-future period considered in CIRCE compared with the end of the century. It is noted above, for example, that the projected changes in mean and extreme temperature for the urban case studies do not exceed the absolute upper range of observed values very frequently (Figs. 5.29 and 5.30). It may be reasonable to assume linearity and stationarity given certain underlying assumptions in the model – which should be made clear to users. In the case of the energy demand model for Athens, for example, it is explicitly assumed that technology use is the same in the control and future periods (Sect. 5.3.3.1). Thus no allowance is made for the potential drive towards a more energy efficient economy or for other socioeconomically-driven trends. A somewhat different approach is taken in the Athens health risk work, where an adaptation or acclimatization factor of 1°C per decade is used in an attempt to account for behavioral and physiological changes.

The case-studies impacts modeling work is focused on the direct effects of climate change rather than on indirect effects and feedbacks. This is, in part, a reflection of the statistical nature of the models used. In the case of the fire risk modeling, for example, no account is taken of potential changes in biomass. Lower biomass is projected for some Mediterranean regions, particularly the south and east, as a result of climate change (Sect. 5.3.2.2) and this could lower fire risk in some regions. In the case of the projected socioeconomic losses for the tourism industry in the Gulf of Gabès (Fig. 5.23), these are based only on the direct effects of temperature on tourist attractiveness. Other impacts of climate change which may also affect the perceptions of tourists, such as beach erosion and loss and water availability, are not reflected in the underlying Daily Climate Tourism Potential Index (Sect. 5.3.3.2). Nor are potential drivers from outside the case-study region considered (Sect. 6.4.1).

Non-climatic influences can be quite difficult to handle in the case-study assessments. They may account for the problems experienced in identifying statistically significant relationships between climate indicators and statistics of wine, wheat and olive oil production for the Apulia case study. A marginally positive correlation was found between summer precipitation and olive oil production, together with weak negative correlations for winter and summer minimum temperature. Wheat production has only weakly positive correlations with summer minimum and maximum temperature and negative correlations with winter and spring precipitation. A negative correlation was found between total wine production and summer and winter maximum temperature, together with a positive correlation for total economic value.

For crop yield, some of the potential problems can be overcome by careful choice of indicator (though this may be restricted by the data available). Yield per hectare, for example, is a better climate-related indicator than total yield for a region: the latter depends on the area planted and is more likely to be influenced by national and European agricultural policy. In the longer-term, however, observed yield per hectare may be affected by changes in factors such as sown variety and use of irrigation and fertilizer. Thus there may be a trade-off between calibrating models over shorter periods (which are less likely to be influenced by non-climatic factors) or over longer periods (in order to capture climate variability and avoid overfitting). In many cases, however, the analyses that can be performed are limited by the length of data available. For the Apulia case-study, for example, only 20 years (1980–2000) of agricultural data were typically available.

Uncertainties due to the limited spatial scale of the climate projections are noted above – particularly in the context of the urban case studies. Ideally, impacts and vulnerability indicators would also be considered at high spatial resolutions. In terms of the health implications of high ozone days, for example, the distribution of vulnerable groups (the young and elderly, those with pre-existing conditions, outdoor workers) may vary spatially across the city and particular 'hot spots' of pollution may occur, for example downwind of industrial areas.

Thus there are uncertainties inherent to all stages of the CIRCE case-studies integrated assessments. The uncertainties in the climate projections are perhaps the easiest to quantify and to demonstrate to stakeholders. Nonetheless, it is important to consider all aspects of uncertainty in the context of adaptation decision making – which is the focus of the next chapter.

References

Beniston M (2004) The 2003 heat wave in Europe: a shape of things to come? An analysis based on Swiss climatological data and model simulations. Geophys Res Lett 31. doi:10.1029/2003GL018857

Beniston M, Diaz HF (2004) The 2003 heat wave as an example of summers in a greenhouse climate? Observations and climate model simulations for Basel, Switzerland. Glob Planet Chang 44:73–81

Beniston M, Stephenson DB, Christensen OB, Ferro CAT, Frei C, Goyette S, Halsnaes K, Holt T, Jylhä K, Koffi B, Palutikof J, Schöll R, Semmler T, Woth K (2007) Future extreme events in European climate: an exploration of regional climate model projections. Clim Chang 81:71–95. doi:10.1007/s10584-006-9226-z

Bigano A, Hamilton JM, Tol RSJ (2008) Climate change and tourism in the Mediterranean. Research unit sustainability and global change FNU-157. Hamburg University and Centre for Marine and Atmospheric Science, Hamburg

Bindoff NL, Willebrand J, Artale V, Cazenave A, Gregory J, Gulev S, Hanawa K, Le Quéré C, Levitus S, Nojiri Y, Shum CK, Talley LD, Unnikrishnan A (2007) Observations: oceanic climate change and sea level. In: Solomon S, Qin D, Manning M, Chen Z, Marquis M, Averyt KB, Tignor M, Miller HL (eds) Climate change 2007: the physical science basis. Contribution of Working Group I to the Fourth Assessment Report of the Intergovernmental Panel on Climate Change. Cambridge University Press, Cambridge/New York

Bloomer BJ, Stehr JW, Piety CA, Salawitch RJ, Dickerson R (2009) Observed relationships of ozone air pollution with temperature and emissions. Geophys Res Lett 36:L09803

Burke EJ, Brown SJ (2008) Evaluating uncertainties in the projection of future drought. J Hydrometeorol 9(2):292–299

Carter TR, Jones RN, Lu X, Bhadwal S, Conde C, Mearns LO, O'Neill BC, Rounsevell MDA, Zurek MB (2007) New assessment methods and the characterisation of future conditions. In: Parry ML, Canziani OF, Palutikof JP, van der Linden PJ, Hanson CE (eds) Climate change 2007: impacts, adaptation and vulnerability. Contribution of working group II to the fourth assessment report of the intergovernmental panel on climate change. Cambridge University Press, Cambridge, pp 133–171

Christensen JH, Hewitson B, Busuioc A, Chen A, Gao X, Held I, Jones R, Kolli RK, Kwon WT, Laprise R, Magaña Rueda V, Mearns L, Menéndez CG, Räisänen J, Rinke A, Sarr A, Whetton P (2007) Regional climate projections. In: Solomon S, Qin D, Manning M, Chen Z, Marquis M, Averyt KB, Tignor M, Miller HL (eds) Climate change 2007: the physical science basis. Contribution of working group I to the fourth assessment report of the intergovernmental panel on climate change. Cambridge University Press, New York/Cambridge, pp 847–944

Court-Picon M, Gadbin-Henry C, Guibal F, Roux M (2004) Dendrometry and morphometry of *Pinus pinea L.* In lower Provence (France): adaptability and variability of provenances. For Ecol Manag 194:319–333

Dessai S (2002) Heat stress and mortality in Lisbon Part II. An assessment of the potential impacts of climate change. Int J Biometeorol 48:37–44

Duvat V, Magnan A (2009) Coastal vulnerability assessment in Djerba (Tunisia). In: Özhan E (ed) Proceedings of the ninth international conference on the Mediterranean coastal environment, MedCoast 2009. Medcoast Foundation, Ankara, pp 355–366

El Sayed Frihy O, Deabes EA, Shereet SM, Abdalla FA (2010) Alexandria-Nile Delta coast, Egypt: update and future projection of relative sea-level rise. Environ Earth Sci 61:253–273. doi:10.1007/s12665-009-0340-x

El-Zein A, Tewtel-Salem M, Nehme G (2004) A time-series analysis of mortality and air temperature in Greater Beirut. Sci Total Environ 330:71–80

Forkel R, Knoche R (2007) Nested regional climate-chemistry simulations for central Europe. Compte Rendus Geosci 339:734–746. doi:10.1016/j.crte.2007.09.018

Founda D, Giannakopoulos C (2009) The exceptionally hot summer of 2007 in Athens, Greece. A typical summer in the future climate. Glob Planet Chang 67:227–236

Giannakopoulos C, Psiloglou V (2006) Trends in energy load demand for Athens, Greece: weather and non-weather related factors. Clim Res 31:97–108

Giannakopoulos C, LeSager P, Bindi M, Moriondo M, Kostopoulou E, Goodess C (2009) Climatic changes and associated impacts in the Mediterranean resulting from a 2°C. Glob Warm Glob Planet Chang 68(3):209–224

Giorgi F, Lionello P (2008) Climate change projections for the Mediterranean region. Glob Planet Chang 63:90–104

Goldammer JG, Mutch RW (2001) Food and Agriculture Organization of the United Nations, Forest Resources Assessment Programme, Global Fire Assessment 1990–2000, Rome, Italy. Available at: http://www.fao.org/docrep/006/ad653e/ad653e02.htm

Good P, Moriondo M, Giannakopoulos C, Bindi M (2008) The meteorological conditions associated with extreme fire risk in Italy and Greece: relevance to climate model studies. Int J Wildland Fire 17(2):155–165

Goubanova K, Li L (2007) Extremes in temperature and precipitation around the Mediterranean basin in an ensemble of future climate scenario simulations. Glob Planet Chang 57:27–42

Harzallah A, Bradai MN, Ben Salem S, Hattour A (2010) In: Agnew M, Goodess C (eds) Biogeophysical and social indicators: coastal case studies information sheet: Gulf of Gabès, Tunisia. Climatic Research Unit, School of Environmental Sciences, University of East Anglia, Norwich, UK. Available from: http://www.cru.uea.ac.uk/projects/circe/Gabes.html

Hawkins E, Sutton R (2009) The potential to narrow uncertainty in regional climate predictions. Bull Am Meteorol Soc 90:1095–1107

Haylock MR, Hofstra N, Klein Tank AMG, Klok EJ, Jones PD, New M (2008) A European daily high-resolution gridded data set of surface temperature and precipitation for 1950–2006. J Geophys Res 113:D20119. doi:10.1029/2008JD010201

Hekkenberg M, Benders RMJ, Moll HC, Schoot Uiterkamp AJM (2009) Indications for a changing electricity demand pattern: the temperature dependence of electricity demand in the Netherlands. Energy Policy 37:1542–1551

Henia L, Alouane T (2007) Le potentiel climato-touristique de la Tunisie. Actes du X Xème Colloque de l'Association de Climatologie, Carthage, 3–8 Septembre 2007

Hohenegger C, Brockhaus P, Schär C (2008) Towards climate simulations at cloud-resolving scales. Meteorol Zeitschrift 17:383–394

Iliadis LS (2004) A decision support system applying an integrated fuzzy model for long-term forest fire risk estimation. Environ Model Softw 20:613–621

ILO (2009) Economically active population estimates and projections, 5th edn. International Labor Organisation, Laboursta. http://laborsta.ilo.org

IPCC (2007) In: Solomon S, Qin D, Manning M, Chen Z, Marquis M, Averyt KB, Tignor M, Miller HL (eds) Contribution of working group I to the fourth assessment report of the intergovernmental panel on climate change. Cambridge University Press, Cambridge, 996 pp

Jacob DJ, Logan JA, Gardner GM, Yevich RM, Spivakovsky CM, Wofsy SC, Sillman S, Prather MJ (1993) Factors regulating Ozone over the United States and its export to the global atmosphere. J Geophys Res 98:14817–14826

Kitoh A, Yatagai A, Alpert P (2008) First super-high-resolution model projection that the ancient "Fertile Crescent" will disappear in this century. Hydrol Res Lett 2:1–4. I: 10.3178/HRL.2.1

Kurz WA, Apps MJ, Stocks BJ, Stocks BJ, Volney JA (1995) Global climate change: disturbance regimes and biospheric feedbacks of temperate and boreal forests. In: Woodwell GM, Mackenzie FT (eds) Biotic feedbacks in the global climatic system. Will the warming feed the warming? Oxford University Press, New York, pp 119–133

Lin C, Jacob DJ, Fiore AM (2001) Trends in exceedances of the ozone air quality standard in the continental United States, 1980–1998. Atmos Environ 35:3217–3228

Lionello P, Elvini E, Nizzero A (2003) Ocean waves and storm surges in the Adriatic Sea: intercomparison between the present and doubled CO_2 climate scenarios. Clim Res 23:217–231

Lionello P, Cogo S, Galati MB, Sanna A (2008) The Mediterranean surface wave climate inferred from future scenario simulations. Glob Planet Chang. doi:10.1016/j.gloplacha.2008.03.004

Lionello P, Galati MB, Elvini E (2010) Extreme storm surge and wind wave climate scenario simulations at the Venetian littoral. Phys Chem Earth. doi:10.1016/j.pce.2010.04.001

Marcos M, Tsimplis MN (2008) Comparison of results of AOGCMs in the Mediterranean Sea during the 21st century. J Geophys Res 113:C12028. doi:10.1029/2008JC004820

May W (2008) Potential future changes in the characteristics of daily precipitation in Europe simulated by the HIRHAM regional climate model. Clim Dyn 30:581–603

McCarthy M, Sanderson M (2011) Urban heat islands: sensitivity of urban temperatures to climate change and heat release in four European Cities. In: Hoornweg D, Frire M, Lee M, Bhada P, Yuen B (eds) Cities and climate change: responding to an urgent Agenda. Publication from the fifth urban research symposium, The World Bank, Washington, DC

MEDD/PNUD (2008) Etude de la vulnérabilité environnementale et Socio-économique du littoral Tunisien face à une élévation accélérée du niveau de la mer due aux changements climatiques et identification d'une stratégie d'adaptation, Phase II, Ministère de l'Environnement et du Développement Durable, 127 pp

Moss RH, Edmonds JA, Hibbard KA, Manning MR, Rose SK, van Vuuren DP, Carter TR, Emori S, Kainuma M, Kram T, Meehl GA, Mitchell JFB, Nakicenovic N, Riahi K, Smith SJ, Stouffer RJ, Thomson AM, Weyant JP, Wilbanks TJ (2010) The next generation of scenarios for climate change research and assessment. Nature 463:747–756. doi:10.1038/nature08823

Sánchez-Arcilla A, Jiménez JA, Valdemoro HI, Gracia V (2008) Implications of climatic change on Spanish Mediterranean Low-lying coasts: the Ebro delta case. J Coast Res 242:306–316. doi:10.2112/07A-0005.1

Schär C, Vidale P, Lüthi D, Frei C, Haberli C, Liniger MA, Appenzeller C (2004) The role of increasing temperature variability in European summer heatwaves. Nature 427:332–336

Semenov MA (2008) Simulation of extreme weather events by a stochastic weather generator. Clim Res 35:203–212

Sheffield J, Wood EF (2008) Projected changes in drought occurrence under future global warming from multi-model, multi-scenario, IPCC AR4 simulations. Clim Dyn 31:79–105

Sousa PM, Trigo RM, Aizpurua P, Nieto R, Gimeno L, Garcia-Herrera R (2011) Trends and extremes of drought indices throughout the 20th century in the Mediterranean. Nat Hazard Earth Syst 11(1):33–51. doi:10.5194/nhess-11-33-2011

Swart R, Biesbroek R (2009) Climate proofing Europe: a comparative analysis of national adaptation strategies. IOP Conf Ser Earth Environ Sci 6(39):392029. doi:10.1088/1755-1307/6/39/392029

UNDP (United Nations Development Programme) (2002) Energy for sustainable development: a policy agenda. UNDP, New York

UNEP/MAP (2005) Mediterranean strategy for sustainable development, a framework for environmental sustainability and shared prosperity, UNEP(DEC)/MED IG.16/. United Nations Environment Programme Mediterranean Action Plan, Athens, 36 pp. www.unepmap.org

van der Linden P, Mitchell JFB (eds) (2009) ENSEMBLES: climate change and its im-pacts: summary of research and results from the ENSEMBLES project. Exeter EX1 3PB. Met Office Hadley Centre, FitzRoy Road, UK, 160 pp. http://ensembles-eu.metoffice.com/docs/Ensembles_final_report_Nov09.pdf

Van Wagner CE (1987) Development and structure of a Canadian forest fire weather index system, Forestry technical report 35. Canadian Forestry Service, Ottawa

Venevsky S, Thonicke K, Sitch S, Cramer W (2002) Simulating fire regimes in human-dominated ecosystems: Iberian Peninsula case study. Glob Chang Biol 8(10):984–998

Viegas DX, Bovio G, Ferreira A, Nosenzo A, Sol B (1999) Comparative study of various methods of fire danger evaluation in southern Europe. Int J Wildland Fire 9:235–246

WAMDI group (Hasselmann S, Hasselmann K, Bauer E, Janssen PAEM, Komen G Bertotti L, Lionello P, Guillaume A, Cardone VC, Greenwood JA, Reistad M, Zambresky L, Ewing JA) (1988) The WAM model – a third generation ocean wave prediction model. J Phys Oceanogr 18:1776–1810

Chapter 6
Synthesis and the Assessment of Adaptation Measures

Clare M. Goodess, Maureen D. Agnew, Debbie Hemming,
Christos Giannakopoulos, Marco Bindi, Camilla Dibari, Hesham El-Askary,
Mutasem El-Fadel, Mamdouh El-Hattab, Mohamed El-Raey, Roberto Ferrise,
José M. Grünzweig, Ali Harzallah, Dina Kanas, Piero Lionello,
César Mösso Aranda, Theib Oweis, Joan Pau Sierra, Marco Reale, Agustín
Sánchez-Arcilla, Mohamed Senouci, Rolf Sommer, and Annalisa Tanzarella

Abstract The final stage of the CIRCE case-studies integrated assessment involved identification and evaluation of the effectiveness of local and regional adaptation options in collaboration with stakeholders, and in the context of wider national adaptation policies and strategies. This stage provides a synthesis of both the case-study work and the wider CIRCE project since it draws on the case-study indicators

---

C.M. Goodess (✉) • M.D. Agnew
Climatic Research Unit, School of Environmental Sciences,
University of East Anglia, UEA, Norwich, UK
e-mail: c.goodess@uea.ac.uk

D. Hemming
Climate Impacts Analysis Team, Met Office Hadley Centre, Exeter, Devon, UK

C. Giannakopoulos
Institute for Environmental Research and Sustainable Development,
National Observatory of Athens, Athens, Greece

M. Bindi • R. Ferrise
Department of Plant, Soil and Environmental Science (DIPSA),
University of Florence, Florence, Italy

Institute for Biometeorology, National Research Council of Italy,
(CNR-IBIMET) Florence, Italy

C. Dibari
Department of Plant, Soil and Environmental Science (DIPSA),
University of Florence , Florence, Italy

H. El-Askary
School of Earth and Environmental Sciences, Schmid College of Science and Technology,
Chapman University, Orange, CA, USA

Department of Environmental Sciences, Faculty of Science,
Alexandria University, Alexandria, Egypt

M. El-Fadel
Faculty of Engineering and Architecture, American University of Beirut, Beirut, Lebanon

A. Navarra and L. Tubiana (eds.), *Regional Assessment of Climate Change in the Mediterranean: Volume 3: Case Studies*, Advances in Global Change Research 52, DOI 10.1007/978-94-007-5769-1_6, © Springer Science+Business Media Dordrecht 2013

for present and future periods together with wider CIRCE work on adaptation options, particularly in the thematic areas of agriculture, forestry and ecosystems, and Mediterranean communities. This synthesis and evaluation links impacts and vulnerability with adaptation, and also benefits strongly from the local stakeholder workshops held towards the end of the project. Lessons learnt and key messages from the CIRCE case studies are presented. While the objectives of the CIRCE case studies have generally been achieved, a number of research gaps and needs remain.

Keywords Climate change • Mediterranean • Integrated assessment • Adaptation • Policy

6.1 Introduction

The projected consequences of global climate change vary considerably across the Mediterranean Basin and across sectors, engendering adaptation a challenge that is specific to physical location and social context. The CIRCE case studies have presented a valuable opportunity to assess adaptation measures specific to the local-to-regional spatial scale, building on community stakeholder knowledge

---

M. El-Hattab
Environmental Studies and Research Institute (ESRI-SADAT),
Minufiya University, Sadat City, Egypt

M. El-Raey
University of Alexandria, Alexandria, Egypt

J.M. Grünzweig • D. Kanas
Institute of Plant Sciences and Genetics in Agriculture, The Robert H. Smith Faculty of Agriculture, Food and Environment, The Hebrew University of Jerusalem, Jerusalem, Israel

A. Harzallah
Institut National des Sciences et Technologies de la Mer – INSTM, Salammbô, Tunisia

P. Lionello
Centro Euro Mediterraneo per i Cambiamenti Climatici, CMCC, Lecce, Italy

Department of Material Science, University of Salento, Lecce, Italy

C. Mösso Aranda • J. Pau Sierra • A. Sánchez-Arcilla
Laboratori d'Enginyeria Marítima (LIM/UPC), Universitat Politécnica de Catalunya
Barcelona, Spain

T. Oweis • R. Sommer
International Center for Agricultural Research in the Dry Areas, ICARDA, Aleppo, Syria

M. Reale • A. Tanzarella
Department of Material Science, University of Salento, Lecce, Italy

M. Senouci
Association Recherche Climat et Environnement – ARCE, Oran, Algeria

(Chap. 2) within the context of the CCIF (Sect. 1.3, Fig. 1.2). The common methodological steps for the case-studies approach involved an initial phase of identifying key research issues at a local or regional scale based on past and present experience (through literature and where possible stakeholder guidance), and defining the scope of the integrated assessment through the development of a case-study CCIF. Among other benefits, these linkage diagrams (e.g., Figs. 3.4 and 5.25) helped to identify specific and critical vulnerabilities relevant to the case studies. In the second assessment stage, a target set of indicators were selected and refined using defined criteria (Chap. 3). Specific climate hazards were analyzed for observed trend and where data permitted, impacts for present and future periods were assessed (Chaps. 4 and 5). The final phase has involved identifying and evaluating the effectiveness of regional adaptation options in collaboration with stakeholders, and in the context of wider national adaptation policies and strategies. In this chapter, we provide an assessment of case-study adaptation activity within the broader structure of the CIRCE project and international framework, and provide a synthesis of the case-study work, including recommendations on research gaps and needs.

In recent years, the international community has increasingly recognized the importance of adaptation as a means of adjusting to and coping with the amount of inevitable climate change that society will experience despite any action that is taken to mitigate these changes through a reduction in greenhouse gases. Adaptation, in the IPCC (2007) has been defined as 'Adjustment in natural or human systems in response to actual or expected climatic stimuli or their effects'. A number of different categories of adaptation have been recognized such as anticipatory (or proactive) adaptation, autonomous (or spontaneous) adaptation, and planned adaptation (based on policy decisions). The concept of adaptive capacity is closely associated with the ability of a system to cope, adjust, or benefit from the consequences of climate change. Building adaptive capacity among Mediterranean countries is relevant to current climate hazards, particularly in the context of extremes such as flooding, drought and heatwaves (Chap. 4), as well as to future climate change (Chap. 5).

6.2 Linking Impacts and Vulnerability with Adaptation

A prerequisite for formulating an appropriate response involves understanding current regional climate hazards, impacts and underlying vulnerability (social and biogeophysical) and appreciating the potential barriers to adaptation. Within the CIRCE case-studies integrating framework (CCIF, Fig. 1.2), societal and environmental vulnerability to climate hazards (sudden onset extremes and longer-term changes) are viewed as a function not only of exposure and sensitivity but also of the capacity of the system to adapt to the consequences of climate change (see Box 1.1 for definitions of key concepts).

6.2.1 Indicators, Thresholds, Coping Range and Adaptation

The CIRCE case studies have adopted an indicator-based approach to the regional assessment of climate change impacts, adaptation and vulnerability. The target indicators constructed for the case studies cover a wide range of social and environmental issues (Chap. 3; Appendix 2) and although by no means comprehensive, collectively build up a picture of current vulnerability to climate hazards in urban, rural and coastal environments. Observed and modeled data have been used to assess for climate sensitivity and to identify any system thresholds (Sect. 3.2.3). These thresholds (or 'tipping points') mark a point of change (non-linearity) in climate sensitivity from a state in which the system is able to adjust to climate change to a state in which the system experiences negative outcomes, such as heat-related deaths. Another type of 'dangerous' threshold defines the point at which system failure occurs (e.g., inundation following the collapse of sea defenses), in this case the response is sudden, the result of a small perturbation (in one or more drivers) crossing a critical point that dramatically alters the fate of the system. In the context of the coastal case studies, a joint probability approach (Stephens and Ramsay 2006) could be used to estimate the risk of future coastal flooding as the joint probability of waves and sea level (in combination with tides), crossing critical thresholds for a given future period.

In some systems, it is possible to define both upper and lower thresholds. For example, in the context of human health, the upper threshold defines the temperature above which heat-related deaths are recorded, while the lower threshold defines the temperature below which cold-related health effects are experienced (see discussion in Sect. 3.5.2). The coping range describes the capacity of the system (in this case human health) to adjust for climate variability, and is defined by the upper and lower thresholds. In these terms, adaptation (such as physiological or social adjustment) can expand the coping range by raising the upper and/or lowering the lower critical thresholds. Thus in hotter climates, the temperature threshold at which heat stress occurs is generally higher than in cooler climates (see Part IV, Sect. 13.2). Thus in the health example, adaptive capacity is related to social and physical vulnerability, and varies between different groups of people, being lower for the elderly, the young, lower income groups, and communities without access to adequate health services.

The additional complexity is that impacts occur as the joint response to the inherent dynamics of climate, biogeophysical and socioeconomic system drivers (or stressors). In the example of water scarcity, the upper threshold is the point at which water resources are unable to satisfy demand. A lower threshold, defines the point at which water resources outstrip storage capacity and lead to flooding. A complete understanding of adaptive capacity involves understanding social and physical stressors and response, including any feedback(s). Improvements to infrastructure (capacity) and water regulation and management can expand the coping range, while inefficiencies in water management, water resource conflicts, inadequate water legislation, and lack of investment in infrastructure can constrain the coping range.

It should be noted that CIRCE climate and impact models have focused on the A1B SRES for future periods up to the middle of the twenty-first century. The environmental and societal consequences will be larger (more damaging) for higher emission scenarios and for future periods around the end of this century. The corollary is that adaptation will be more challenging for more extreme climate and more damaging consequences.

Information regarding the future timing, frequency and magnitude of exceeding thresholds has particular application in stakeholder planning decisions and adaptation preparedness. As a threshold approaches, the stakeholder can take appropriate or optimal action aimed at preventing or reducing negative outcomes. While knowledge of future threshold exceedances has considerable stakeholder value (Sect. 3.2.3), there are major challenges in providing this information due to the large uncertainties that exist (Sect. 5.4). It is also possible that the threshold responses modeled using observed data, may not hold true in future periods, especially in the case of rapid climate change. With data adequacy, extremes can be calculated, but the uncertainties remain large. For example in the Athens case study, the probability of ozone exceedance days (max 8-h average 60 ppb) has been calculated for two future periods (2021–2050 using three CIRCE climate models; 2071–2100 using ENSEMBLES output) and compared to observed values (1990–1999). By the middle of the present century there could be up to eight additional ozone exceedance days per year, and by the end of the century there could be 27 additional days relative to the baseline period (Fig. 5.15, Sect. 5.3.3.1). However, the uncertainties remain to be more clearly quantified (projections are based on only one socioeconomic scenario), and although ozone pollution has been closely linked to adverse effects, the severity of the community health risk and interactions with other environmental variables is unclear.

6.2.2 Vulnerability, Adaptation and Adaptive Capacity

Key vulnerability indicators have been identified for each case study (Table 6.1). In addition to climate change, vulnerability is affected by changes in social systems. Population growth, for example, drives change in the demand for resources (e.g., water, energy, land) and in environmental dynamics (atmospheric, marine and freshwater pollution; soil erosion, over-exploitation of resources), and increases exposure to environmental hazards (such as coastal flood). Common to many of the Mediterranean case studies are vulnerabilities in water availability, this is particularly the case for urban and rural case studies where additional non-climate drivers (such as population growth or increasing competition for resources) has reduced water availability. Across generic case-study types, pollution (atmospheric in the case of Athens and Alexandria; freshwater in the case of Beirut) is a key vulnerability indicator for urban case studies. For the rural case studies, soil quality (soil erosion, salinization of soil, or soil nutrients) and land use change are common aspects of vulnerability and reflect the importance of agriculture to the rural economy. Many

Table 6.1 Key vulnerability indicators identified for the CIRCE integrated case studies

Generic case study	Case study[a]	Vulnerability indicators identified
Urban	AT:	Fire Weather Index; ozone exceedance days; Urban Heat Island index; area of peri-urban forest; environmental management; access to health services; population growth; demographics (>65 or <5 years)
	BE:	Water resource management, groundwater abstraction, freshwater pollution, water availability, salt-water intrusion, access to clean water, use of water vendors, access to health services, population growth; demographics (>65 or <5 years)
	AL:	Sea-level; coastal protection expenses; population/infrastructure at risk from flooding; water availability; soil salinity, atmospheric pollution; population growth; demographics (>65 or <5 years); access to health services
Rural	TU:	Water availability; water use; water resource management; land use/area of agricultural land; fire weather index; soil erosion; agricultural contribution to GDP/employment; population growth
	AP:	Water availability; water use; land use/area of agricultural land; soil erosion; drought index; agricultural contribution to GDP/employment in agriculture; population growth
	TH:	Water availability (especially groundwater); water resource management; salinization; freshwater pollution; area of agricultural land; agricultural practices; agricultural contribution to GDP/employment in agriculture; population growth; regional development/sustainability
	JF:	Water availability; water use; land use changes; environmental management/protection; tree stand density/age; soil nutrients; agricultural practices; sector (agriculture/forestry/tourism) employment/contribution to GDP; population growth; regional development/sustainability
Coastal	VA:	Sea level; wave height; freshwater outflows; marine pollution/invasive species; salt-water intrusion; loss of agricultural land; sector contribution to GDP/employment; water use; population growth in the coastal zone; infrastructure at risk from flooding; extent of the Marine Terrestrial Public Domain
	OR:	Pollution discharge into the sea; invasive species; sea level; coastal protection; nature reservations; fluvial/estuarine pollution; sector contribution to GDP/employment; population growth
	GA:	Sea level; swell height; coastal protection; land loss by erosion/salinization; saline intrusion of coastal aquifers; marine pollution; invasive marine species; tourism seasonal shift index; sector contribution to GDP/employment; population growth
	WN:	Sea-level; water availability; salt-water intrusion; atmospheric pollution; population growth; land use at risk from coastal flooding; wildfires; coastal protection; marine pollution, invasive marine species

[a] See Appendix 1 for case-study acronyms

of the coastal vulnerability indicators are common across case-study sites, and include sea-level rise, level of coastal protection, pollution of marine and estuarine waters, and invasive marine species.

Quantifiable vulnerability indicators (Table 5.3) fulfill the established selection criteria (Sect. 3.2.1) and are particularly relevant for working with stakeholders in the case-study regions. However, long-term records for many of these vulnerability indicators are not available for the case studies (e.g., Beirut in general; soil quality for the rural case studies; environmental monitoring in coastal areas). Nonetheless for these more qualitative vulnerability indicators there may be useful information on general trends or one-off surveys. For some of these indicators, data are available at a national level, but are not publicly available at a regional or local level (e.g., in the case of Greater Athens, the Judean Foothills, Tel Hadya, and Alexandria), and very few regional projections for social indicators are available.

Adaptation strategies for the case studies should be focused around the identified critical environmental and social vulnerabilities. Societal response should be aimed at reducing the damaging consequences of climate variability and change through integrated and sustainable management of vulnerable resources and communities. General recommendations include, improving environmental management and protection of water resources, air quality, coastal defenses, soils, forests and conservation areas; flora and fauna. A series of regional sector adaptations are proposed as part of the individual case studies integrated framework, and the wider national adaptation strategy (Sects. 6.3.1 and 6.4.2). The Judean Foothills case-study conceptual framework, for example, makes specific reference to 'Response and adaptation' for reducing vulnerability in rural settlements (Fig. 3.4).

The ability to respond is clearly as important as the adaptation strategy. Previous work on adaptation (e.g., Füssel and Klein 2006; Ionescu et al. 2009; EEA 2010) has identified some more 'theoretical' integrative indicators of vulnerability and, in particular, adaptive capacity, at a national level. These national indicators of vulnerability and adaptive capacity reflect the importance of non-climatic stressors, and are useful for illustrating the large variability among the countries of the Mediterranean (Table 6.2). In particular, there are some large differences between the European Mediterranean countries (lower population growth rate; and higher rank in human development index, literacy rate, democracy index and in the corruption perception index), and WANA (West Asia North Africa) countries in terms of social vulnerability indices. It is also likely that some social stresses such as population growth will continue to drive change for a large part of the present century (Sect. 5.3.4.2; Fig. 5.24).

Adaptive capacity is considered in the wider CIRCE project as a function of social capacity, economic capacity, technological eco-efficiency (e.g., agricultural innovation), natural capital, and climate capital. These five components have been used to compute a national adaptive capacity index (ACI) for nine countries of the Mediterranean (Part IV, Sect. 11.2; Fig. 11.1). For the six case-study countries in common (Spain, Italy, Syria, Egypt, Tunisia and Algeria), the ACI is generally higher among the European countries of the Mediterranean (particularly Italy) and least for the non-European countries (particularly Syria). Economic and social

Table 6.2 National vulnerability and adaptive capacity indicators for the case-studies

Case-study countries	HDI	Literacy rate	Population growth rate	Democracy index	CPI
Spain	15	97.9	1.52	8.34	6.7
Italy	18	98.9	0.33	7.73	5.2
Greece	25	97.1	0.23	8.13	4.6
Israel	27	97.1	1.66	7.28	6.1
Syria	107	83.1	2.52	2.36	2.4
Lebanon	83	89.6	1.05	5.82	–
Egypt	123	66.4	1.76	3.90	2.9
Tunisia	98	77.7	1.08	3.06	4.2
Algeria	104	75.4	1.51	3.17	3.0

HDI is the Human Development Index 2007 (UNDP 2009); literacy rate is for adults aged 15 and above 2007 (UNESCO 2009); the population growth is annual, % for the years 2000–2005 (WRI); Democracy Index ranges from 0 to 10, for the year 2007, from The Economist; CPI is the Corruption Perception Index ranging from 0 to 10, from Transparency International (an indication for the adherence to formal agreements). Adapted from RACCM Part III, Sect. 2.6, Table 2.2

capacity contribute substantially to the higher ACI values for European countries, technological eco-efficiency is also generally higher for European countries (although relatively lower for Spain), and is particularly low for Syria (Part IV, Sect. 11.2, Fig. 11.2). This assessment is particularly useful for the development of national-level policies (Sect. 6.4.2) aimed at redressing lower-level ACI components. Political unrest and upheaval (occurring at the time of writing in several case-study countries) can increase human insecurity and, at least in the short term, reduce social capacity for adaptation and undermine community resilience. However, in the longer term political changes have the potential to increase economic and social adaptive capacity through a process of post-conflict reconstruction (Sect. 6.4.1).

There is limited baseline knowledge of sub-national vulnerability and adaptive capacity in the public domain. The CIRCE case studies set of quantitative and qualitative indicators focus on aspects of climate change impacts and vulnerability of specific relevance to each region (Chap. 3). The 'top-down' and 'bottom-up' approach (Chap. 1) to climate change impact and vulnerability assessment (Chaps. 4 and 5) aimed to make use of all relevant knowledge sources as a basis for identifying and evaluating local and regional adaptation measures in collaboration with the local stakeholder community (Chap. 2).

6.3 Adaptation Measures at Local and Regional Scales

A key theme from the earlier parts and chapters of the RACCM is the need for integration at a variety of scales and sectors, and between scientists and stakeholders allowing a sharing of tools, 'best practice', data and information. The case studies approach views climate as only one of several drivers, and the outcomes involve complex interactions between the dynamics of biogeophysical and social systems (CCIF, Fig. 1.2). Consequently, development of adaptation strategies should include consideration of both physical and social factors. Mainstreaming (Huq et al. 2003; Klein et al. 2005; McEvoy et al. 2008) adaptation measures involves solutions that strengthen capacity ('no regrets options') to deal with both current and future problems. An integrated approach can also present a more attractive solution to stakeholders, since the evaluation of adaptation strategies should assess not only the negative impacts avoided, but the indirect benefits to health and well being, food and energy security, and employment opportunities.

6.3.1 The CIRCE Stakeholder Perspective on Adaptation

Preparation for climate change should be an inclusive endeavor involving the participation of scientists, decision-makers and all levels of society to formulate and implement effective strategies. The CIRCE case studies focus on identifying evidence-based adaptation measures that help communities adjust to climate change at a regional and local level. Regional stakeholder workshops were organized in the case-study locations (Table 6.3) as a means of sharing knowledge about region-specific climate hazards (past, present, and future), and identifying and evaluating adaptation measures in the context of critical environmental and social vulnerabilities (Chap. 2). The extent to which it has been possible to glean stakeholder views on adaptation, and level of awareness and interest in climate change varied from one workshop to another. The remainder of this section outlines the key adaptation messages and other relevant issues emerging from each of the stakeholder workshops and discussions.

6.3.1.1 Athens Case Study Workshop

The Athens urban case study assessed adaptation options through a series of interviews with stakeholders and scientific experts. Among decision makers, the current focus was adapting to current climate hazards, such as flooding and heatwaves. There was also an awareness of the need to take a longer-term perspective. A range of adaptation strategies was suggested across several sectors. Health protection in Athens is adopting a proactive stance in anticipation of an increasing risk of vector-borne disease (such as malaria, yellow fever, dengue, and West Nile virus), that involves vector monitoring, larvae control, and the use of mosquito repellent. In addition, specific health advisories are issued in advance of an imminent heat wave, and stadiums are deployed as designated cooling zones. Several changes in the forestry sector have been observed, an expansion of Aleppo pine to higher elevations in the peri-urban forests of Parnitha, more insect activity, and increasing risk of forest fire. Additional financial investment in the Greek Fire Service is required to take into consideration climate change, and improve forest management.

Table 6.3 Format and approach for the regional stakeholder workshops

Case study[a]	Location (number of participants)	Workshop language(s)	Material used	Workshop approach
AT	Athens (five expert stakeholders)	English and Greek	Athens briefing notes and information sheets, perception questionnaire	Series of interviews conducted with expert scientists and stakeholders
AL/WN	Alexandria (45 participants of which 15 were stakeholders)	English and Arabic	Alexandria and West Nile Delta briefing notes, presentations, work groups, perception questionnaire, open discussion	Plenary sessions consisting of case-study presentations, open discussion, working group discussions, workshop evaluation
TU	Dept of Plant, Soil and Environmental Science, University of Florence (30)	Italian	Workshop flyer, background material from key speakers, 'World café' methodology, perception questionnaire prepared in stakeholder collaboration. All material was translated into Italian and circulated in advance of the workshop	Plenary sessions (CIRCE project and case-study presentations, 4 invited speakers presentations); 'World Café' (2 subgroups: agriculture; tourism); plenary feedback session; workshop evaluation by questionnaire
AP	Lecce (60 participants of which 9 were stakeholders)	Italian	Background case-study information; perception questionnaire	Invited speaker presentations; roundtable discussion; contributions by stakeholders, presentation of the 'Playdecide' game by the CIRCE communication office[b]
TH	Seminars at ICARDA and Aleppo university	English and Arabic	Models, supplemental irrigation data and scenarios for adaptation	The meetings included presentation of the approaches, modeling results, field trials and preliminary outputs. Discussions on the potential impacts and possible adaptation strategies
JF	The Robert H. Smith Faculty of Agriculture, Food and Environment, the Hebrew University of Jerusalem (15)	Hebrew	Judean Foothills briefing notes and information sheets translated into Hebrew; workshop goals	Plenary sessions on case-study indicators, open discussion of impacts; open discussion on adaptation to climate change

GA	Tunis (26). A smaller workshop (~10 participants) was held in Tunis in 2008 to present case-study objectives and discuss data availability/use	French	A questionnaire on vulnerability and adaptation perception was distributed during the workshop	Plenary sessions with 2–3 key presentations following by discussion, workshop feedback
OR	Oran (25–30)	French and Arabic	Translated case-study briefing notes and information sheets, climate projections, special reports and photographs, perception questionnaire (in advance)	First workshop: lead by case-study team, 2–3 key presentations, focus groups, and plenary feedback sessions. Second workshop: stakeholder organized and led
VA	Cartographic Institute of Catalonia; Deltanet International Conference	Spanish	Information sheets and published articles	Case-study presentations followed by discussion fora on adaptation options

The Beirut urban case study collaborated with stakeholders using a series of informal meetings and seminars in the place of more formal workshops
[a]See Appendix 1 for case-study acronyms
[b]CIRCE Playdecide is a 'participatory method' (www.playdecide.eu), a game for discussing climate change issues and impacts in the Mediterranean area

In the tourism sector, there is a realization that to meet the challenges of climate change, services need to be redefined and diversified to reduce detrimental environment impacts. This will require public and private sector collaboration. In the context of urban planning the need for specific coordinated adaptation plans is recognized. These will involve the promotion of green spaces, bioclimatic building, and creating a coastal buffer zone free from buildings and infrastructure. The Ministry of Environment is implementing a plan to increase the energy efficiency of low-income housing, municipal and government buildings.

6.3.1.2 Beirut Case-Study Dialogue

Water scarcity and water security (particularly in relation to human health and well being) has been a particular focus of the Beirut urban case study. Through previous research and in consultation with a wide range of local stakeholders a set of alternative adaptation strategies has been assessed in terms of their applicability and feasibility (need and cost). Included in this assessment is a consideration of the potential challenges to implementing each strategy (Table 6.4). The potential solutions cover a wide range of response from no action, through to hard engineering 'grey' options (dam building; reverse osmosis desalinization technologies), 'green' infrastructure (such as rain harvesting through the installation of roof collecting systems, and reclamation of wastewater), and a flexible combination of different strategies to meet different (potable/non-potable) needs ('smart' approach to adaptation). However, current 10-year national water resource plans rely more heavily on 'grey' engineering interventions than exploiting the 'greener' alternatives. Specific health adaptation options for the Greater Beirut Area have been considered in parallel with water management strategies. Appropriate health measures include increasing disease resilience through improved sanitation and hygiene, improved health monitoring programs (including the use of diagnostic tests and disease prevention technology), and enhanced public education and training of health care professionals. In addition, the development of disaster preparedness plans and urban heat warning systems could increase resilience to current and future climate hazards and reduce premature mortality.

6.3.1.3 Tuscany Case-Study Stakeholder Workshop

In the agricultural sector in Tuscany, decision-making operates within a relatively short time horizon, and identified risks are associated with short-term variability in climate. Stakeholders were very aware of the influence of climate variability on agricultural activities, and helped clarify regional climate thresholds in the production of some crops (vineyard, olive and cereal). In particular, the explicit risk of late frosts to vineyard and olive production was discussed, and two specific climate hazards for cereals were better understood. First, prolonged rainfall that delays sowing between February and March. Second, higher temperatures coupled with

Table 6.4 Water adaptation strategies considered in the Beirut urban case study

Adaptation strategies	Applicability/feasibility	Challenges
Relying on conventional water supply sources[a]	Optimal only in the absence of constraints in the use of available sources (groundwater and surface water)[a]	Supply sources are threatened by climate change. Limited resources for expansion plans (Awali-Beirut conveyor, Bisri and Damour dams, treatment plants, distribution networks)
Combining seawater desalination and wastewater reclamation[a]	Necessary when socio-environmental policy is considered[a]	Desalination cost is high with respect to supply charges. Social/religious constraints on wastewater reuse
Applying rainwater harvesting (roof tapping)[a]	Becomes beneficial if the water charging scheme is modified to allow for more return, along with socio-environmental constraints[a]	Investment in installation of collection and distribution networks in buildings
Expanding both conventional and non-conventional resources[a]	Mandatory when tourist demand increases[a]	All of the above
Using treated surface water and desalinated seawater for urban potable uses, and untreated groundwater and reclaimed wastewater for non-potable uses[b]	Wastewater reuse for non-potable supply is cost effective, even when the cost of non-potable water networks is included[b]	Social/religious constraints on wastewater reuse
Using desalinated seawater for potable supply, reclaimed water for groundwater recharge. Groundwater can then be used (untreated) to provide non-potable supply[b]	When taking into consideration the anticipated increase in demand and decrease in water availability due to climate change, this is a approach for meeting potable and non-potable demands[b]	Wastewater treatment plant (secondary) may not be able to meet the stringent requirements for groundwater recharge (Ghadir and Bourj-Hammoud)

[a]Yamout and El-Fadel (2005) consider wet season only
[b]Ray et al. (2010) consider wet and dry seasons

reduced precipitation around anthesis (flowering) and grain-filling phases. In contrast to the limited time perspective of the agricultural sector, there is increasing interest in long-term planning in the regional tourism sector, and a sustainability initiative based in the *Regional Observatory of Tourism* is showing increasing interest in climate change initiatives. Diversification of economic activity in response to the inter-annual vagaries of farmer income is increasingly common in rural areas. The provision of farmhouse accommodation for countryside tourism supplements annual incomes and strengthens the capacity to cope with external environmental and social pressures. However, both agricultural yields and tourist inflows are influenced by short- and long-term climate variability. In recent years, the tourist market has shown a trend away from the traditional long summer seaside holiday to a larger annual number of 'short breaks' spread between a variety of destinations. It is not clear whether climate has a major role in this process, or whether cultural trends is the key driver of change.

Two 'World Café' thematic tables (Sect. 2.3) on agriculture and tourism were organized in the stakeholder workshop to discuss the capacity and strategies for coping with current and future climate impacts on rural activities in Tuscany. The tourist sector at the regional level is perceived to have a reasonable degree of flexibility and adaptive capacity. However, stakeholders recommend that clear channels of cooperation should develop between regional tiers of adaptation, and adaptation at the micro-scale tier of individuals and organizations. It was suggested that to develop effective adaptation plans requires further detailed understanding of the regional tourist response to climate variability and change for 'shoulder' seasons (April-June, September and October). Consideration should be given not just to the impacts of climate on tourist flows, but also to the impacts of climate on tourist resources (e.g., water availability) and the interaction between these sectors (e.g., the rising tourist demand for water resulting from the increasing provision of swimming pools on Tuscan farms).

Several specific adaptation strategies were assessed during stakeholder discussion. For cereals, the use of later varieties was perceived a feasible strategy for coping with rain delays to sowing, and genetically improving crops to adapt to changing climate conditions was viewed an essential activity. To increase the resilience of vineyards to late frosts, frost-resistant rootstock could be introduced. However, in the context of developing adaptation strategies for rural regions, the whole market supply chain should be considered rather than simply targeting appropriate strategies for agricultural practices. As a policy recommendation, stakeholders emphasized the need to invest in technical training of farmers to better cope with climate hazards. It was also suggested that identifying 'tipping points' could be a helpful process for determining whether a management strategy is able to meet its objectives in the face of climate change. For regional rural tourism, the diversification, and active support of more sustainable practices that provide high quality services with lower environmental impact was considered an optimal long-term strategy.

6.3.1.4 Apulia Case-Study Stakeholder Workshop

The Apulia case-study team struggled to engage regional stakeholders, administrators and decision makers throughout the course of the project. There was a reluctance to collaborate with scientists until dedicated (regionally focused) results had become available. Historically, environmental problems have generally not been given high priority by politicians in Italy and this presents a potential communication barrier. Perhaps as a consequence, the participation of stakeholders in the Apulia workshop was low (17%; 9 people) relative to scientists and students. There is therefore a pressing need to improve communication of key climate change issues in this region, and encourage a greater use of participative methods.

Despite these challenges, in a perception questionnaire, workshop participants listed a variety of climate hazards which they perceived to be important (including intense precipitation, heat waves, and conditions favoring pollutant episodes). Similarly, a diverse range of climate impacts were recorded as important (such as, melting glaciers and sea-level rise, drought, irregular water availability, agricultural management and food availability), demonstrating a high level of awareness. The adaptation measures that most participants identified for reducing harmful impacts of climate change were high-level overarching strategies, such as reducing greenhouse gas emissions (i.e., mitigation rather than adaptation), and were largely in keeping with an ethos of sustainable development (e.g., bio-fuel production, reforestation, use of renewable energy sources, use of energy-efficient technologies, and promotion of local products). The adaptation measures identified centered on water-agricultural linkages and included optimal use of irrigation water, creating irrigation basins in old quarries for dry periods, optimized agricultural practices, diversification of water sources and reducing leaks in water supply. Participants were asked to select relevant timeframes for evaluating climate change and adaptation strategies. The majority of stakeholders chose long-term 16 plus years (56%) and mid-term 3–15 year time frames (44%) while a smaller proportion of participants chose an evaluation time frame that was short-term (22%) or multi-decadal, more than 30 years (22%). A key message from the Apulia workshop was that the stakeholders desired to be well informed and to be consulted in policy decision-making processes for adapting to climate change.

6.3.1.5 The Judean Foothills Case-Study Stakeholder Workshop

There was a high level of awareness of climate change among the stakeholder community of the Judean Foothills rural case study, and there was anecdotal evidence that the consequences of climate change are already being 'observed'. Stakeholders perceive that a run of dry years has been weakening and damaging plants, altering vegetation patterns, and that this has been brought about by a shift in the climate regime. Extreme heat has lowered the quantity and quality of fodder crops, fruit and milk production, while there has been negligible yield from almond trees due to the lack of low temperatures failing to break dormancy. A wide-range of adaptation options were considered in stakeholder discussion. These were primarily targeted at increasing sustainability, such as conserving water resources through domestic price increases, altered systems of gardening, increase in desalinization and 'grey' water use.

Adaptation to current climate hazards is in evidence, such as afforestation plans by the Jewish National Fund using germplasm (the basic genetic material for a plant) for drought-resistant trees, e.g., *Cupressus sempervirens* (Italian Cypress), *Pinus halepensis* (Aleppo pine), *Pistacia atlantica* (Mount Atlas mastic tree) and local Acacia species. There are additional plans to reduce fire risk vulnerability through 'best practice' responses, such as livestock grazing control, and buffer lines (a practice developed in Spain). Efforts to increase ecosystem resilience to climate change and conserve natural resources (e.g., through the creation of traditional vineyard terraces) are perceived to lead to positive 'knock-on' effects in the tourism sector by enhancing destination attractiveness (e.g., wine tourism promotion following the *Wine Routes of Israel*). A set of adaptation options has been developed for the agricultural sector that covers a full range of response types. These comprise for example, farm-level reactive responses such as using wheat crops for fodder instead of grain in years for which grain harvest is not profitable, 'green' proactive responses to revert to traditional (and biblical) less water intensive crops (such as olives, almonds and grapes), and technological proactive adaptation (such as developing new varieties of deciduous fruit trees with less winter chilling required to break dormancy).

6.3.1.6 The Tel Hadya Case-Study Seminars

The Tel Hadya case study focused on the use of supplemental irrigation of rain-fed agricultural systems as an adaptation to climate change. The analysis focused on wheat, but the methodologies are also applicable to other cereals and legumes. Stakeholders were interested in understanding the impacts of climate change on crop growth duration and patterns, including possible changes in the timing and duration of drought spells and precipitation totals during the growing season. Uncertainties in these projections caused some frustration for most of the stakeholders (who included farmers and the local irrigation water authority). Research and extension stakeholders were most interested and wanted to follow-up the results. Awareness of the potential impacts of climate change on agriculture was generally not high among the stakeholders. For many stakeholders it was hard to distinguish between natural climate variability and human-induced climate change. There was also a challenge in making the stakeholders aware that it is not just water issues that need to be analyzed, but all aspects of the agricultural sector. Thus it is recommended that an awareness campaign targeted at policy makers, in particular, is created.

6.3.1.7 Alexandria and West Nile Delta Stakeholder Workshop

In the Alexandria and West Nile stakeholder workshop, there was particular interest in understanding the potential impacts of climate change for specific sub regions. Delegates were invited to consider what might be the most effective and affordable adaptation measures that could be implemented in the region in the context of their organization. Suggestions included reducing greenhouse gas emissions (i.e., mitigation), cooperation with (and support from) government departments such as the Ministry of Health and coordinating the adaptation response among organizations. With response to the specific challenges of sea-level rise for the low-lying delta region, stakeholders thought that the most important action was to develop and implement an integrated coastal zone management plan for the Nile Delta coast to protect low-lying areas and agricultural land. In the context of Lake Mariut (south of Alexandria), it was accepted that there should be an integrated approach to finding an acceptable solution to the rising demand for water and negative consequences for the lake ecology and the local fishing industry. During the workshop there was some evidence of capacity building in action, with several of the delegates suggesting that it had been particularly important for the different organizations represented to share experience, and that all parts of society (from specialists and executive authorities to the general public) needed to increase their awareness of the potentially negative consequences of climate change, and to actively participate in the process of responding to climate change.

6.3.1.8 Gulf of Gabès Case-Study Stakeholder Workshop

Stakeholders in the Gulf of Gabès seem very well informed (through the media, workshops and conferences) on climate change issues at a general level, and claim to have already 'observed' some physical changes such as the damaging effects of sea-level rise. The participants also expressed a greater desire for technical and scientific, socio-economic and political information. This high level of awareness may in part be due to past and recent experiences of extreme events on key economic sectors in Tunisia. The perceived climate-related risks for stakeholder activities in the region include sea-level rise, inundation and coastal erosion, desertification, drought, salinization and loss of marine biodiversity. A key message from the workshop was that adaptation plans should target the tourism and fishing sectors, and littoral and marine ecosystems. An emphasis was also placed on the need for concerted adaptation actions at all levels (local, regional and national) with complementary mitigation measures such as the use of renewable energy. Improved awareness and management were viewed as important in the implementation of adaptation plans. Actions directed at reducing social stressors (e.g., overfishing) could reduce ecosystem vulnerability to current climate hazards and future climate change. Some adaptation plans are already in evidence (at the level of managers, officials and executives) with mid-term (3–15 years) or long-term (16 plus years) time frames. Based on adaptation strategy reports with specific action plans for coastal zones, sea-level rise projections are now accounted for in all new coastal constructions.

The CIRCE project was initially seen by stakeholders as yet one more project dealing with regional climate change impacts. However, communication and presentation of project results were decisive in stakeholders recognizing that CIRCE constitutes a fundamental step in the understanding of climate change impacts and adaptation. Two key specific features of the CIRCE case-study work were crucial. The first is its integrative nature, combining climate, biogeophysical and economic aspects (e.g., climate warming, sea-level rise, alien species, employment losses) and economic sectors (e.g., fishing tourism) with very simple and easy-to-use indicators. The second is the case-study approach being as close as possible to stakeholder expectations. Interest in this *Regional Assessment of Climate Change in the Mediterranean* was noticeably greater by the end of the local workshop.

6.3.1.9 Gulf of Oran Meetings with Scientists, Civil Society and Decision Makers

The Gulf of Oran case study was the focus of a scientific workshop organized as part of an international seminar on 'Biodiversity and Coastal Ecosystems' and also of a meeting with a local NGO specializing in marine environmental protection in the Bay of Oran. The latter meeting was also an opportunity to publicize the 'blue clubs' created in schools and colleges in partnership with the Oran education authorities. Data on waste categories affecting the coast of Oran collected by these clubs will be analyzed by the CIRCE local case-study team. Discussions were also held with the Directorate of the Environment and other local decision makers. As a result, a working group has been set up, initially involving the local case-study team and the Directorate, aiming to undertake further studies of the CIRCE climate projections and their implications for impacts and vulnerability in the region. These will be undertaken in partnership with two other regional initiatives: the launch of the 'Regional Plan of Land Management in Western Algeria' and development of the 'National Climate Plan'.

6.3.1.10 Gulf of Valencia-Catalan Case-Study Stakeholder Workshop

Through voluntary agreements, industry in the Catalan area has progressively reduced emissions of greenhouse gases. A new framework program for climate change mitigation has begun to stimulate emissions reduction action across other primary economic sectors. This framework aims to support organizations leading the way in climate action, raise public awareness of the need for climate action, and share experience and best practice. However currently, management of climate risks and adaptation strategies remains at a generally low state of awareness across industrial and civil sectors of society. The general consensus among stakeholders was that present policies are more focused on mitigation than adaptation. Given recognition of mutual climate impacts across multiple sectors, a more cross-sector approach to adaptation planning is required that supports greater regional government responsibility. Key adaptation options highlighted for the wider region include promoting the use of sustainable and more energy efficient irrigation and agricultural systems, encouraging use of renewable sources of energy, and improved integrated plans for the coastal zone that take into consideration other system feedbacks. The most realistic adaptation option for the coastal zone was thought to involve artificially maintaining the present shoreline through beach nourishment and the use of flexible engineering structures to partially limit shoreline movement. The importance of stakeholder participation in decision making and developing adaptation plans was highlighted as a means of avoiding conflict and encouraging cooperation. To be successful, consultation with stakeholders and wider society should take place in the early stages of adaptation projects.

6.3.2 Consolidating the Case-Study Information on Adaptation

Convening the stakeholder workshops has contributed to building regional level adaptive capacity in terms of providing and exchanging information and knowledge, and in raising awareness. The desire for more detailed, local information was particularly evident in the feedback from the Alexandria/West Nile and Apulia workshops. In addition, the existing use of and demand for meteorological and severe weather forecasts apparent in Apulia, highlights the need for adapting to current climate variability and hazards.

By their very nature the case studies cannot provide a comprehensive assessment of adaptation, and the workshops were further constrained in number, in scope, and in available time for discussion. Despite these obstacles, key adaptation messages are beginning to emerge from the case studies, and can be considered in a theoretical and research-focused adaptation framework (such as European Commission 2009; EEA 2010), distinguishing between types of adaption response (Table 6.5). For example, adaptation strategies may involve: (1) Behavioral, managerial and policy responses ('soft' or 'smart' adaptation measures); (2) Ecosystem-based responses ('green' adaptation measures), and (3) Technological responses ('grey' adaptation measures). 'Grey' adaptation options comprise physical 'concrete' technological interventions, such as hard engineering solutions to increase the resilience of buildings and infrastructure. The 'green' adaptation approach increases ecosystem resilience and is seen as an example of 'synergy and no-regret' options. The first option, 'soft' solutions is a behavioral, managerial and policy approach, that tends to be flexible, incorporates safety margins, and accounts for other actions being implemented (e.g., Wilby and Dessai 2010).

In terms of timing, adaptation responses can be further subdivided into proactive planned or anticipatory measures, and autonomous or reactive measures. For example, within the rural case studies a proactive managerial response describes the use of different cereal cultivars to increase resilience to future climate extremes, while autonomous strategies include a diversification of tourist services at the farm level to supplement irregular or diminished income. 'Green' ecological and technological-based adaptation strategies can also be considered as adaptation channels (or facilitators) that can be widened to include 'smart' financial, political and social-based networks for strengthening, assisting and implementing adaptation plans. 'Green' adaptation responses are generally low-cost options and examples tend to focus on the rural CIRCE case studies and include increasing plant resilience to heat or drought through the use of new or different cultivars, and improving water use efficiency. These sustainable strategies include a combination of high level policy actions promoting reforestation and afforestation (Apulia, Judean Foothills), the use of renewables and energy-efficient technologies, and local-scale strategies such as

Table 6.5 Examples of case-study adaptation options summarized within the framework of the EU White Paper on Adaptation (2009)

Approach	Case-study examples[a]
'Grey'	Creating irrigation basins in old quarries (AP)
	Construction of desalinization plants (JF; BE)
	All new coastal construction incorporate sea-level rise projections (GA)
	Artificial beach nourishment combined with flexible engineering structures to partially limit shoreline movement (VA)
	Dam building schemes, e.g., Bisri and Damour; water transfer schemes (BE)
'Green'	Increase energy efficiency of buildings (AT)
	Sustainable rural tourism practices (TU)
	Different varieties of crops resilient to changing climate conditions (TU)
	Use of germplasm for drought-resistant trees (JF)
	Reconstruction of traditional vineyard terraces (JF)
	Less water intensive crops – almonds, grape, olives (JF)
	Varieties of deciduous fruit tree requiring less winter chilling to break dormancy (JF)
	Frost-resistant rootstock (TU)
	Bio-fuel production, reforestation, local product use (AP)
	Reduce water supply leaks (AP)
	Increased use of 'grey' water for cultivation (JF)
	Supplemental irrigation schemes (TH)
	Actions to reduce overfishing and increase marine ecosystem resilience (GA)
	Wastewater reclamation; rainwater harvesting (roof tapping) (TH)
'Soft' or 'Smart'	Health preparedness plans for increasing social resilience during heat wave events, e.g., designation of 'cooling zones' (AT)
	Coordinated urban development plans which promote 'green' urban spaces (AT)
	Improved communication and integration between regional and national adaptation strategies (TU)
	Technical training of farm management (TU)
	Optimizing irrigation practices (AP)
	Use of livestock grazing and buffer zones to reduce fire risk (JF)
	Government support in the provision of coordinating adaptive response among organizations (AL/WN)
	Developing and implementing an integrated coastal zone management plan for the Nile Delta (AL/WN)
	Concerted multi-level (local, regional, national) actions (GA)
	Integrated cross-sector approach to adaptation planning (VA)
	Improved and integrated plans for the coastal zone that take into consideration other system feedbacks (VA)
	Systematic environmental monitoring and data collection (BE)
	Heat wave warning systems to improve resilience and reduce disease burden (BE)

[a]See Appendix 1 for case-study acronyms

reducing water supply leaks (Apulia), wastewater reclamation and rain-water harvesting schemes (Tel Hadya). Investment in 'green' research and development strategies include genetic improvement of crops (Tuscany, Judean Foothills), and 'smart' identification of 'tipping points' in the response to climate, concomitant with investment in technical training for farm-level adaptation support (Tuscany). 'Grey' infrastructure options are less frequently considered among the case studies. Innovative high-tech engineering solutions generally have higher costs, research and training requirements, than 'greener' alternatives. Examples include, the construction of desalinization plants (Judean Foothills, Beirut) and creating irrigation basins in disused quarries (Apulia); coastal engineering structure to limit shoreline movement and protect against coastal flooding (Gulfs of Valencia and Gabès), and large-scale dam building and water transfer schemes (Beirut).

The outcomes of several of the stakeholder workshops (including the Gulf of Gabès, Alexandria and the West Nile Delta) support a coordinated 'smart' adaptive response among government departments and organizations. This can be found for example, in the implementation of an integrated coastal zone management plan (ICZM) to protect the Nile Delta coast and in coordinated community strategies to cope with diminishing freshwater resources and conflicts of interest. Stakeholders in the Gulf of Gabès emphasized the need for concerted anticipatory actions on multiple spatial levels that complement 'green' adaptation measures such as the use of renewable energy and decreasing ecosystem vulnerability (e.g., by reducing overfishing and controlling the spread of invasive marine species).

Work in the wider CIRCE project has begun the process of identifying types of adaptation options that are likely to be more important or effective for specific sectors and themes. The following thematic paragraphs highlight activities relevant to the case studies, and further outline the scope and context for adaptation.

6.3.2.1 Agriculture, Forestry and Ecosystems

Mediterranean ecosystems may naturally develop physiological mechanisms (e.g., advance in flowering, leaf unfolding, and delay in leaf fall) in response to climate change. Some phenological changes have already been observed and documented. In addition, there may be a species shift towards higher latitude and elevation sites. Socio-economic factors will also be important in the response to climate change. Therefore, integrated adaptive ecosystem management is required that aims to maintain or increase biodiversity, and involves strategies of water conservation, grazing control, and afforestation with tree species best adapted to future climate conditions (this is already being implemented in the Judean Foothills, see Sect. 6.3.1.5). Focused community and regional agendas are needed to overcome barriers to adaptation, such as competing societal demands for land, wood products, and ecosystem services (Part III, Sect. 5.9).

Several adaptation strategies have been proposed to cope with the negative effects of climate change on agricultural yields (e.g., advanced sowing date, using drought-resistant varieties) that could be easily adopted by farmers on an individual basis.

In the Judean Foothills, adaptation measures in agriculture involve a wide range of issues, such as the use of recycled wastewater instead of the limited supplies of freshwater, a shift to more drought- and heat-adapted crops (often the more traditional crops or varieties), breeding of better adapted varieties, and management strategies for coping with increased pests and disease. However in Tuscany, a switch to other varieties of crops may be unfeasible due to farmers' reliance on industry to sell their products. Individual farmers have limited freedom to choose varieties of wheat that are more resistant to climate change. Paradoxically, the use of 'better' adapted varieties of wheat could be frustrated through lack of interest by the wheat industry because these varieties do not meet required standards of quantity or quality. However, a more coordinated response by farmers would increase their bargaining power with industry, eliminating the risk of production fragmentation with the establishment of small niche markets.

Mediterranean wildfires adaptation strategies are focused on reducing fire hazards and increasing ecosystem resilience to accommodate changed climate conditions especially in areas of high vulnerability (Part III, Table 6.1). Resilience can be strengthened through concerted planting of 'keystone' species and land restoration schemes. These schemes aim to increase species diversity, improve water conservation and availability, and prioritize prevention and post-fire restoration actions which promote woodland regeneration and halt ecosystem degradation. The use of spatial decision support systems (SDSSs) can facilitate forest management of fire vulnerable areas. CIRCE has tested a GIS-based SDSS in Ayora, Valencia to evaluate a variety of fire-control scenarios and develop a composite risk map showing priority areas for management and protection. A key recommendation of this work is the use of woodlands to fragment large fire-prone areas and effectively reduce fire size, while also promoting biodiversity (Part III, Chap. 6).

6.3.2.2 Mediterranean Communities

Societal responses to real and perceived climate change hazards comprise collective and individual actions within water, energy, tourism and health sectors. In the CIRCE project (Part IV, Sect. 11.5.2), a Water Availability and Policy Assessment (WAPA) model has been used to identify optimal water management given regional projections of climate change (including the CIRCE climate models). For the energy sector, options for cost-effective 'green' adaptation strategies include the proposal to bring North African countries into the European emissions market and the import of low-cost carbon-free electricity (solar, wind) from the North African countries. Three alternative mitigation scenarios were considered within the CIRCE project (Part IV, Chap. 14); EU-alone (assumes present mitigation policies), international concerted effort (based on more ambitious international pledges for emission reduction), and CSP (concentrated solar power) scenario (international concerted effort plus the deployment of CSP plants in North Africa). Using computable general equilibrium modeled projections of GDP, welfare and employment for 2030, world activity is reduced for all three scenarios. However, the outcome for North Africa is actually improved for the CSP scenario compared to the EU-alone scenario (Part IV, Table 14.26), particularly for GDP and welfare in Egypt (approximately +4% and +7% for the CSP scenario compared to no change for the EU alone scenario).

Mediterranean tourism resorts require adaptation strategies to cope with the dual hazards of additional heat stress and sea-level rise (Part IV, Chap. 15). Suggested strategies include product diversification at the resort level to emphasize the favorable qualities of Mediterranean destinations relative to their northern European counterparts, efforts to promote tourism in the shoulder seasons (autumn and spring) when temperatures are less extreme (e.g., in the island of Djerba, the Gulf of Gabès). To be effective, adaptive strategies need to be resort-specific and begin with a thorough vulnerability assessment in collaboration with local decision makers (Part IV, Sect. 15.5.2). This is particularly the case where environmental regulation is limited, and stakeholders are a key driver of development. Developing cross-sector adaptation strategies and solutions is important for increasing infrastructure resilience and building adaptation capacity. The Gulf of Gabès case study has developed a GIS stakeholder tool to facilitate cross-sector planning. It is available on CD-ROM in two languages (English and French), and includes regional information on coastline, bathymetry, main cities, harbors, atmospheric parameters, oceanic parameters, coastal indicators, and biodiversity indicators.

Health preparedness actions and policies should be targeted at protecting the most vulnerably in society (those with limited access to health care, the young, the elderly, and those with pre-existing illness). Within the CIRCE project, a pilot health adaptation training workshop was held in Athens, June 2008. Strategies designed to promote better health care and improve well being present a 'win-win' or 'no regrets' solution. This has been highlighted in the Beirut urban case study which has projected future health impacts for heat stress and water- and food-borne disease. It is suggested that sufficient 'soft' behavioral adaptation (such as improved disease monitoring, hygiene and sanitation) could increase the overall social capacity of the population to resist disease (such as diarrhea, hepatitis, and typhoid) while simultaneously reducing health insecurity in the face of climate change.

6.3.2.3 Integrating Adaptation Options Across Scales and Sectors

To make informed adaptation decisions, stakeholders and politicians require as complete an understanding of the uncertainties in climate projections, future risks and potential consequences, as possible. While some broad changes in climate and impacts are relatively certain at the Mediterranean level, there are much greater uncertainties at the local and regional scale represented by the case studies (Sect. 5.4). The level of detail and confidence required by stakeholders also depends on the types of decisions being made, i.e., detailed information regarding water availability at a sub-catchment level, or broader and more general information regarding regional warming. Acknowledging the large uncertainties that exist in future projections of climate, social and biogeophysical systems, approaches such as Robust Decision Making (RDM) may be useful for designing adaptation plans (Lempert et al. 2006;

see also Part IV, Sect. 12.4.2). The CIRCE results are largely based on the A1B SRES scenario; however, this scenario is no more likely than any other. The RDM approach attempts to look for strategies that are robust across a range of climate and socio-economic scenarios, rather than trying to pin-point an optimal strategy for a given climate future (Part IV, Sect. 12.4.2).

At the spatial scale of countries and regions, adaptation strategies are required to respond, for example, to existing water shortages and future projections of diminished water resources. Solutions may emerge through the mainstreaming of climate information into existing policies. However, it may also be necessary to develop new economic and legislative frameworks (e.g., for the management of coasts or water resources) to facilitate and invest in large-scale long-term adaptation. At subregional scales, examples include urban flood planning and the creation of 'green' city spaces (see Sect. 6.3.1.1). Organization level examples of adaptation to climate change include changes in the (re)location and design of infrastructure and facilities, altered provision of goods and services, use of innovative technology, and insurance against negative impacts of climate extremes. At this spatial level of adaptation, there are potentially more opportunities for mainstreaming responses into existing operations, development and business plans, and 'no regrets' strategies (such as those targeted at natural hazards and climate extremes) may be particularly beneficial in early stages of adaptation. At a subsidiary scale, the development of building codes for climate change is an example of single building level response. Public perception surveys suggest that individuals or small groups will take both anticipatory and reactive action to minimize the perceived negative consequences of climate change and current climate variability. Despite the generally low probability of occurrence, individuals pay particular consideration to extreme climate hazards such as floods.

Adaptive capacity and response is a function of a complex array of processes operating at multiple space and time scales and is extremely context specific. Adaptive capacity at a local scale is constrained and/or driven by larger-scale economic and political processes. For example at a farm level, adaptive capacity depends both on access to resources required for crop production, and crop prices determined by macro-scale economic processes (Adger et al. 2005). As nations and regions of the Mediterranean devise adaptation plans, it will be important to consider multiple spatial and temporal scales to develop effective strategies and avoid maladaptation. A short-term perspective response to higher temperatures, such as the installation of air conditioning equipment in buildings, could have negative consequences in the long-term. Cross-scale linkages need to be developed between small and large-scale adaptation projects, with the creation of overarching national and international facilitating and guiding structures.

The CIRCE project has drawn attention to the need for a much more integrated approach to adaptation in the Mediterranean region (Part IV, Sects. 10.2, 11.2.1 and 12.2.1.3). Climate change cannot be adapted to in isolation. Other non-climate drivers and their integrated impacts (including the impacts of policy intervention) and climate system feedbacks, need to be considered in a future world. Integrated approaches (e.g., see Table 6.5) should consider a wide range of interventions, comprising multiple adaptation channels (from ecological to financial-based approaches), and multiple scales (from urban district or farm level projects to international scale projects). Discussion of adaptation in the case-study stakeholder workshops concurs with this recommendation. Adaptation to climate change is a long-term iterative process involving preparatory, development, implementing, testing (monitoring and sharing of best practice) and refining stages. To provide initial impetus, early adaptation strategies may have a greater reliance on 'soft' adaptation channels and 'no regrets' options, while subsequent more advanced and robust strategies will show a greater diversity in adaptation options.

6.4 Moving Beyond the Case Studies

Work in the case studies has focused on what is happening and what may happen in the case-study regions as a consequence of the dynamics in climate and social systems. However, there is also a need to consider events, stresses, and policy beyond the case-study level (i.e., at national, Mediterranean-wide, and global scales) that have an influence on processes at the regional level, particularly in the context of vulnerability and adaptation decision making.

6.4.1 Changes, Impacts and Events Outside the Case-Study Regions

6.4.1.1 Supply and Demand Issues (for Energy, Water, Food)

The effects of climate change on an individual region cannot be considered in isolation, but need to be considered in a global context. Global supply and demand issues (e.g., for energy, water, food products and raw materials) can have implications for both adaptation and mitigation in the case studies. For example, effects of crop failures elsewhere will be reflected in global market prices and ultimately impact on agricultural activity in the target regions. Traditional Mediterranean crops such as durum wheat, olives, and grapes are also grown in China, the USA, Russia, and northern India. A major reduction in durum wheat yield globally, could drive up prices regionally and cause consumers and/or farmers to look for alternative products. Thus, regional economic impacts are not solely a response to climate-related yield changes, but are modified by international trade which contributes to the comparative economic advantages (and disadvantages) of regions.

6.4.1.2 Global Effects on Regional Tourism

Global effects are also important for the tourism market, with changes in the relative attractiveness of resorts globally altering the attractiveness of resorts within the case study region. Thus when projecting changes to tourism flow resulting from climate change, it is particularly important to widen the spatial horizon of interest. It can be as important to consider climate change impacts in the source regions for Mediterranean tourism as it is to consider climate change impacts in Mediterranean tourist regions. CIRCE research has suggested, for example, that higher temperatures (particularly in summer) will have a negative impact for the Mediterranean on outbound tourism from Britain (Part IV, Chap. 15).

6.4.1.3 Migration

The wider CIRCE project has focused on the role of climate in international retirement migration (IRM) from northern Europe (and in particular the United Kingdom) to the Mediterranean (Part IV, Chap. 16). Climate is found to be an important factor in the decision to migrate in retirement, but it is also affected by cultural expectations, distance and travel costs. The attraction of the Mediterranean for retirement migration from northern latitude countries is driven by the greater warmth (neither too hot in summer nor too cold in winter) and sunshine of the countries bordering the Mediterranean relative to other countries globally. Tuscany, a case-study region, has historically been a preferred destination for IRM, although cultural factors have been found to be more important than climate (King and Patterson 1998). Changes in comparative attractiveness of the host region in relation to climate change could have economic implications, but these are complex since IRM can have negative (e.g., health and welfare costs) and positive (e.g., property investment and the creation of new jobs to meet increased demand for services) consequences.

IRM is only one component of migration. There are economic and political migrant fluxes from rural areas to urban areas or coasts, from southern to northern Mediterranean regions, and from countries outside the Mediterranean region (e.g., sub-Saharan Africa) into the Mediterranean region. The EU project EACH-FOR (Afifi 2009) examined reasons for forced migration in the Nile valley and delta. The focus was on the significance of environmental degradation as a motivating factor for migration. Migrants and expert witnesses perceived poverty and unemployment to be key motivating factors in the decision to migrate. However, these economic factors could be partially attributed to environmental decline and in turn, to climate variability and change. Thus migration is the outcome of complex multi-level coupled processes operating between human and environmental systems, with a range of potentially positive and negative consequences for the host region or city. Deserving further consideration for Mediterranean regions is the view (e.g., McLeman and Smit 2006) that migration can be considered as an adaptation response (rather than a failure of adaptation) allowing redistribution of climate-affected populations and diffusing resource conflicts.

6.4.1.4 Governance Issues

Local and national governance issues influence both vulnerability to climate change and adaptive capacity. Moser and Ekstrom (2010) suggest that through lack of guidance, governance can weaken the capacity and motivation to make adaptation decisions. Austerity measures implemented in many EU countries in response to periods of economic downturn may indirectly undermine plans for adaptation where budgets supporting climate change action are trimmed. Countries which have recently experienced conflict are some of the nations most vulnerable to climate change (Brooks et al. 2005). Abrupt changes in governance (such as those which have recently occurred in North Africa) can introduce uncertainty regarding climate policy implications, at least in transitional and early stages (Sect. 6.2.2). However, at medium and longer time-scales the changes in governance in North Africa could facilitate the establishment of new and more efficient adaptation strategies and actions covering a larger number of ecosystems. Integrated governance mechanisms can strengthen adaptive capacity. For example, Engle and Lemos (2010) show that integrated water resource governance mechanisms are positively associated with adaptive capacity, but there remain tensions and tradeoffs between some variables, namely 'equality of decision making' and 'knowledge available'. The governance framework largely determines the outcome of negotiation between competing actors. Multi-level governance networks promoting sustainable development, equitable natural resource management, and decision-making structures that accommodate risk uncertainty and greater social participation, are key to strengthening resilience and meeting the challenges of climate change (Adger et al. 2005; Nelson et al. 2008; Eriksen et al. 2011).

6.4.2 Adaptation Policy

International cooperation to understand and meet the challenges of climate change has included the work of the IPCC, and the Conference of the Parties of the UNFCC. The European Commission (2007a) released a communication that set out a proposal to limit global warming to a minimum of 2 °C above pre-industrial temperatures in an attempt to prevent irreversible impacts of global climate change. To meet this target, it was proposed that the EU should commit to a unilateral 'greenhouse' gas emissions reduction of 20% by 2020 (relative to 1990 levels), increasing to a 30% cut as part of a future global agreement. In the long term, it was stipulated that developing countries need to reduce emissions by 50% relative to 1990 levels by 2050. In preparing to meet the challenges of climate change, the European Commission (2007b) has published a Green Paper on adaptation options in Europe, and a White Paper on moving towards a *European Framework for Action* (European Commission 2009).

Some European member and non-Annex I countries have prepared (or started to prepare) *National Climate Change Adaptation Strategies* (for a recent discussion see Biesbroek et al. 2010). However, adaptation is still not on the agenda of some Mediterranean countries. Among the CIRCE case-study countries (Table 6.6), only Spain has adopted National Adaptation Strategies (NAS), *The Spanish National Climate Change Adaptation Plan*, (PNACC: 2006). Several other countries (Greece,

Table 6.6 Summary of adaptation activities by case-study country

Case-study country	UNFCCC National Communications (most recent year of submission)	National Adaptation Strategy (NAS)	Other key adaptation projects and activities
Spain	5th (2009)	Yes	PNACC – *The National Plan for the Adaptation to Climate Change* (2006), administrative adaptation framework, Spanish Ministry of Environment
Italy	2nd (2010)		Italian government institutions (2005 onwards): Strategic Programme for Sustainable Development and Climate Change. Preparation activities for a National Adaptation Strategy are at an early stage (Sgobbi and Carraro 2008; WHO-APAT 2007)
Greece	5th (2010)	Yes, in prep.	A National Action Plan for Combating Desertification (2001). Draft Programme for the Management of Water Resources in Greece (2003). National Strategic Reference Framework 2007: elaboration of a National Strategy for the Adaptation to Climate Change
Israel	2nd (2010)	Yes, in prep.	Ministry of Environmental Protection 2006: steering committee on adaptation to climate change. *Israel's Adaptation to Global Climate Change* (2008). Israel Ministry of Agriculture & University of California at Davis: Biotechnology for Agriculture in Saline Environments project. Jewish National Fund: afforestation using drought resistant species. Blaustein Institute for Desert Research: design of climate sensitive housing
Syria	1st (2010)		UNDP-GEF: Enabling Activities for the Preparation of Syria's Initial National Communication to the UNFCCC: 2007–2009
Lebanon	1st (1999)		UNDP EEG and GEF: Enabling Activities for the Preparation of Lebanon's Second National Communication (SNC) to the UNFCCC 2007–2011
Egypt	2nd (2010)		Adaptation to the Impacts of Sea Level Rise in the Nile Delta Coastal Zone www.idrc.org/ccaa; UNDP-GEF: adaptation of the Nile Delta to climate change through integrated coastal zone management. 2009–2014
Tunisia	1st (2001)	Yes	Ministry of Agriculture & water Resources–GTZ (2007): *National strategy for agriculture and ecosystems adaptation to climate change*; Ministry of Environment & Sustainable Development–UNDP-GEF (2008): *Study of the environmental and socioeconomic vulnerability of the Tunisian littoral to sea level rise due to climate change*; Ministry of Environment & Sustainable Development–GTZ (2010): *Tourism and Climate change in Tunisia, impacts and national strategy for adaptation*
Algeria	2nd (2010)		UNDP-GEF: 3-year vulnerability and adaptation project for the Maghreb region (Algeria, Morocco, Tunis a) –building adaptive capacity

Israel and Tunisia) have initiated preparatory work for developing a national adaptation framework, and while NAS have not formally been developed in the countries of the southern and eastern Mediterranean, adaptation tends to have a relatively prominent role. At the 16th session of the UNFCC Conference of the Parties (COP16, Cancun, Mexico, 2010), the *Cancun Adaptation Framework* was developed to facilitate action on adaptation through international cooperation on capacity building and vulnerability reduction. In addition, a position paper was published on *Climate Change in Coastal Zones of the Mediterranean* (Travers et al. 2010) through the UNEP Priority Actions Programme which reviews the current status of adaptation initiatives in the Mediterranean region to develop a set of recommendations for the Mediterranean region implemented using the ICZM (Integrated Coastal Zone Management) protocol.

The challenge to develop policy that responds to a changing but uncertain future is exemplified in the water sector. Indicator assessments at a sub-national level provide a potential means of identifying critical areas of vulnerability to water scarcity and resource conflict in urban, rural and coastal areas (Chap. 4), and highlight areas that require targeted strategies to increase adaptive capacity in the management of water resources (see Sect. 6.3.1, and Part IV, Chaps. 11 and 12). The EU Water Framework Directive (WFD) is an example of a legal intervention that places emphasis on understanding the environmental status of water bodies before abstraction licenses are granted. Countries such as Greece are designing and implementing regional adaptation strategies that are compatible with the much broader WFD. Re-use of urban water, water pricing mechanisms, and upgrading of network infrastructure are strategies for increasing water use efficiency and reliability. Particularly in the context of water scarcity in Mediterranean countries, high level adaptations are likely to require effort across several sectors and will be influenced by activities in other sectors (e.g., agriculture and ecological conservation schemes promoting fog collection, and rainwater harvesting; see Table 6.4).

A mismatch of scale exists between the general concerns of adaptation policy and the more specific concerns of adaptation practitioners. While adaptation policy considers national and political borders, climate change and its consequences have no consideration for such socially imposed boundaries. Stakeholders tend to be concerned with the methods of implementing adaptation strategies at a more local level (Brown et al. 2011). To be efficient, adaptation strategies must therefore understand critical areas of vulnerability, and the degree of sensitivity and exposure to climate change at a local and regional level. The case studies have attempted to fill this gap, seeking to inform policy makers and regional stakeholders of hazards, vulnerabilities and sensitivities at a scale relevant to their activities. However, due to resource and time constraints, they do not and cannot provide a fully comprehensive assessment of climate change vulnerability and adaptation.

Adaptation policies for the Mediterranean should not be separated from other policies such as development, planning and environmental protection. The evidence of the local and regional case-study impact and vulnerability assessments suggests that without concerted and integrated effort climate change is likely to accentuate the current pressures and inequalities that exist in the countries bordering the

Mediterranean Sea. The development of adaptation plans to build resilience and reduce vulnerability, and their successful implementation depends on the cooperation of policy makers, local and regional stakeholders, and individuals.

6.5 Conclusions: Lessons Learnt and Key Messages from the CIRCE Case Studies

The rigorous common framework (the CCIF – Sect. 1.3) developed to assess the integrated risks of climate change has been vital in working towards the key objectives of the CIRCE case studies (Sect. 1.2). At the heart of the CCIF are the climate, biogeophysical and socioeconomic indicators (Chap. 3) and the diagrams constructed for each set of generic case studies (urban: Fig. 4.2; rural: Fig. 4.6; and, coastal: Fig. 4.11) and for each individual case study (for examples, see Figs. 3.4 and 5.25) summarizing the linkages between hazard, impact and vulnerability. These diagrams demonstrate that climate dynamics and hazards are only one aspect that needs to be considered in integrated assessments and clearly reflect the importance of biogeophysical and social dynamics (impacts) and vulnerability (see also Fig. 1.2).

Despite some constraints of time and financial resources, and data availability, the objectives of the CIRCE case studies have generally been achieved. In the end, however, it was not possible to explicitly consider mitigation measures and, as discussed below, the bottom-up work with stakeholders on adaptation measures was also somewhat constrained.

The CCIF provides a flexible framework which can be implemented at different levels of detail including different levels of quantitative analysis. Thus it would be readily transferable to other case studies and regions. Even where only limited quantitative analysis was possible, the CCIF was useful in framing thinking by researchers and local decision makers. It provides a mechanism for identifying key issues and inter-linkages and also for identifying research needs and gaps (Sect. 6.6).

Some common difficulties were, however, experienced in implementing the CCIF in practice. A detailed set of criteria was developed for case-study indicator selection (Sect. 3.2.1), but ensuring that these criteria were met was not always easy. The difficulties relating to the identification, construction and use of indicators are highlighted throughout this volume and are listed here:

- Limited data availability and quality (Sects. 3.2.4 and 5.4)
- Difficulties in identifying and separating climatic and non-climatic influences on the indicators (Sects. 3.2.4 and 5.4)
- Difficulties in distinguishing between impacts and vulnerability in the biogeophysical and socioeconomic indicators (Sect. 3.2.4)
- Difficulties in identifying thresholds (particularly in human/social systems) and using them consistently (Sects. 3.2.3 and 6.2.1)
- Difficulties in identifying coping ranges (Sects. 3.2.3 and 6.2.1)

The extent to which the selected indicators could be quantified, especially for future periods, varies considerably (Table 5.3). The indicators which were most straightforward to quantify are temperature and precipitation-based climate indicators. Observed meteorological station data were, however, lacking or of doubtful reliability in some cases. There are, for example, concerns about the homogeneity of the Beirut weather data particularly during the civil war period (Sect. 5.4). Analysis of the CIRCE climate model simulations was constrained by time and the nature of being a downstream user.

Both the observed and projected trends towards higher temperatures and more frequent and more intense high-temperature extremes appear to be robust (Table 6.7). A number of the case studies show acceleration in warming since the mid-1970s and a continuation of this trend into the future. The consistency of observed and projected temperature trends makes it appropriate to use experienced high-temperature events, such as the hot summer of 2007 in the Eastern Mediterranean (Box 5.1), as analogues for a future warmer world, and for exploration of impacts, vulnerability and potential adaptation measures. Despite the robustness in the projected direction of temperature change, both the CIRCE coupled climate model ensemble and the earlier ENSEMBLES uncoupled RCM ensemble reveal uncertainty in the magnitude of the future changes (Sect. 5.4). The availability of the latter ensemble is helpful in providing supporting evidence from a larger ensemble (more than 20 runs in total) than available from CIRCE (a maximum of six, with many of the currently-completed case-study analyses based on only four or fewer of these runs). Other published studies provide further supporting evidence, at least for the direction of temperature change, although many are focused on the end of the twenty-first century when the climate change signal is greater than for the 2021–2050 scenario period considered in CIRCE (Sect. 5.4).

For precipitation and other variables such as wind, waves, storms and sea level (Sect. 3.2.4), issues of data availability and reliability, compounded by their greater inter-annual and spatial variability, make it difficult to identify robust changes – particularly in the case of observations. In general, there is a lack of statistically significant observed trends in mean precipitation identified for the case studies, and it is even harder to make statements about changes (either observed or projected) in precipitation extremes. Despite these issues, there is a robust and consistent decrease in mean precipitation projected by the CIRCE models for all the case studies, except for Alexandria and the West Nile Delta (Table 6.7). This is consistent with evidence from other Mediterranean studies of projected increases in dry spell duration and meteorological/agricultural drought (Sect. 5.2).

Care is needed in making statements about future changes in extremes such as heavy precipitation and flooding, wind and storm surges in the case-study regions. It is also important that the uncertainties in projections are communicated to users and stakeholders. The size of the CIRCE ensembles precludes taking a probabilistic approach such as developed in ENSEMBLES (van der Linden and Mitchell 2009), but it is nonetheless important to present the ensemble range as well as the ensemble-mean change (see Figs. 5.1, 5.2, 5.3, 5.4, 5.5, 5.6, 5.7, 5.8 and 5.9). It is also important that users and stakeholders are aware of the nature and extent of climate

Table 6.7 Summary of observed (last few decades) and projected (2021–2050) changes in mean temperature (Mean T), high temperature extremes (High T) and mean precipitation (Mean P) in the case-study regions

	Observed			Projected (2021–2050)		
	Mean T	High T	Mean P	Mean T	High T	Mean P
Athens	↑	↑	(↑)	↑	↑	↓
Alexandria	↑	↑	(↑)	↑	↑	↔
Beirut	↑	↑	↔	↑	↑	↓
Tuscany	↑	↑	↓	↑	↑	↓
Apulia	↑	↑	↓	↑	↑	↓
Judean Foothills	↑	↑	↔	↑	↑	(↓)
Tel Hadya	↑	–	↔	↑	↑	↓
Gulf of Valencia	↑	–	↑/↔	↑	↑	↓
Gulf of Oran	↑	–	↓	↑	↑	↓
Gulf of Gabès	↑	↑	↔	↑	↑	↓
West Nile Delta	↑	↑	(↑)	↑	↑	↔

↑: observed/projected increase, ↓: observed/projected decrease, ↔: no trend/change identified. Brackets () indicate very small/uncertain trends/changes. –: not analyzed

projection information underlying impacts assessments – particularly where the number of climate models used is limited (Sect. 5.4). Within the time and resource constraints of the CIRCE case-study work, it has not been possible to work with all CIRCE climate models, although analysis of this output is ongoing at the time of writing.

In terms of the analysis of both climate hazard indicators and biogeophysical and socioeconomic impacts and vulnerability indicators, some common issues arise. The first concerns the spatial scale at which it was possible to undertake analyses. The climate projections over land have a spatial resolution of 25 km grid boxes at most, whereas finer-scale mesoscale modeling would be required to fully resolve effects such as that of urban morphology on the urban heat island and the occurrence of flash flooding in urban areas (Sect. 5.4). At the same time, data for construction of many of the biogeophysical and social indicators is only available at the country level rather than at the finer scales more appropriate for the development of adaptation strategies (Sect. 6.3.2). This is likely to be more of an issue for larger and geographically diverse countries such as Spain and Italy than for some of the smaller case-study countries. In other respects, the geographical focus of the case studies might be considered too restricted since it does not explicitly encompass potential drivers from outside the region such as tourist demand and migration (Sect. 6.4.1).

It is also important to consider the seasonality and timing of observed and projected changes. The susceptibility of plants to extreme weather events, for example, may be very dependent on phenological stage. Consideration of seasonal cycles and life cycles may also be important with respect to other impacts. Fish stocks in the Gulf of Gabès, for example, are in part dependent on seasonal fish spawning areas (which are themselves sensitive to changes in SST and other characteristics of the marine environment). Sequences and combinations of climatic hazards (such as repeated flooding, combined coastal and riverine flooding, drought followed by flood) are also potentially important, but have not been explored as part of the CIRCE case-study work and published information on such events is very limited.

When presenting observed and projected changes in climate, there can be a tendency to focus on the winter and summer seasons – whereas some of the major impacts may occur in the transitional seasons. Summer precipitation changes are not so relevant or interesting when there is no or very little precipitation as in many parts of the Mediterranean (and percentage changes can be very misleading in such circumstances). In terms of water availability, it is important to consider the main periods of groundwater recharge and also crop growth. Thus extended winter (October to March) precipitation was used as a climate indicator for some of the case studies (Fig. 5.8).

There is a need to look at the total picture and the annual balance of change. This is especially important where the relationship between climate and impact is bimodal or V-shaped – as in the case of temperature-related deaths and energy demand (Fig. 4.3) in Athens. There are currently more cold-related winter deaths in Athens than heat-related summer deaths. The case-study work explored potential future changes in heat-related deaths (Sect. 5.3.3.1, Figs. 5.16 and 5.17) but did not consider how the balance of cold/heat-related deaths might change. This is, however, being done in the case of heat-related mortality in Beirut (Sect. 5.3.4.2). Work with the DCTPI index in the Gulf of Gabès shows a balance between summer losses and autumn/winter gains in climate conditions conducive for tourism – with overall a negative balance projected for the future (Sect. 5.3.3.2).

As already noted above, it has proven difficult to separate out climatic and non-climatic influences on indicators. This issue is discussed in Sects. 3.2.4 and 5.4 and some additional specific examples of where this has been problematic are listed here:

- Shoreline modification and erosion in the Gulf of Gabès
- Observed decline in Camarote prawn catches in the Gulf of Gabès
- Observed trends in forest and peri-urban fires in Athens, Tuscany and Apulia
- Land-use changes in Tuscany and Apulia (Sect. 4.3.3)
- Increased energy consumption in Athens (which is directly related to increasing GDP and greater installation/use of air conditioning).

In terms of projecting future impacts and vulnerability, the CIRCE case-study modeling work focused on the direct impacts of the projected climate changes, rather than being able to incorporate indirect effects and feedbacks (see Sect. 5.4).

One of the key objectives of the CIRCE case studies was to use a risk-based bottom-up iterative approach based on regional stakeholder workshops (as well as the top-down approach based on indicators) to identify and explore potential adaptation and mitigation options (Sect. 1.2). Financial constraints meant that it was only possible to hold one round of workshops in each of the case-study regions (Sect. 6.3.1); although informal contact and communication was maintained throughout the project in many cases (Chap. 2). At the start of the project, it was anticipated that the level of stakeholder involvement would be at the level of 'participation', i.e., acting to ensure processes and results are relevant (Sect. 2.2 and Fig. 2.1), but in practice, given that the workshops were held towards the end of the project, the formal level of involvement was perhaps closer to that of 'consultation' in some cases. Both the research teams and the stakeholders (as evidenced by some of the workshop comments – Sect. 6.3.1) would have welcomed the opportunity for more extensive and intensive contact and interaction. In addition, resource and time constraints meant that it was not possible to test some of the more complex tools for stakeholder interaction developed in the project (Sect. 2.3). The World Café approach was, however, used successfully in Tuscany and the 'Playdecide' game developed by the CIRCE communications office was presented at the Apulia workshop (Table 6.3). Despite these limitations, stakeholders were involved in the identification of relevant indicators and, in particular, provided valuable input on the development of adaptation strategies during the local workshops (Sect. 6.3). The CIRCE experience has also provided the basis for good practice guidelines on facilitating stakeholder involvement in research (Box 2.1). Two key messages with respect to adaptation emerged from the local stakeholder workshops:

- There should be concerted adaptation strategies implemented at all levels from individual/community, to regional, national and transnational levels.
- Linking adaptation action into sustainability initiatives is a potentially useful means of reducing climate vulnerability in a preliminary stage.

The stakeholder input on adaptation was particularly valuable since adaptation options were not modeled in any of the CIRCE case-study impacts assessments (with the exception of the adaptation or acclimatization factor included in the Athens statistical model of high-temperature health impacts – Sect. 5.3.3.1). The CCIF, nonetheless, provides a fully appropriate and flexible framework for the exploration of adaptation strategies (Sect. 6.2).

As well as access to regional stakeholders, one of the selection criteria for the CIRCE case studies was their vulnerability to climate change (Sect. 1.1). Thus it is not surprising that natural and human systems in all 11 case studies were found to be vulnerable to current climate variability and change as well as to social dynamics or drivers (Chap. 4). The climate projections of increases in mean and extreme high temperature and decreases in precipitation (Table 6.7) are considered to be robust, although there is uncertainty with regards to the magnitude of change (Sects. 5.2 and 5.4). They indicate that all case studies will experience continuing and increasing vulnerability to climate change in the absence of mitigation or adaptation. Projections for other extreme weather events, such as heavy precipitation and flooding, are highly uncertain, but any increase in such events would further increase vulnerability. Very few examples have been identified of where climate change might be beneficial for the case studies or where there might be opposing or compensating changes. At the same time, social dynamics and drivers such as population growth (at least in the short term and in the southern Mediterranean) are likely to further increase vulnerability (Sect. 5.3.4.2). The local interplay and balance of factors will reflect the spatial complexities of the Mediterranean natural and human

6 Synthesis and the Assessment of Adaptation Measures

environments. In terms of broad geographical patterns, however, while the largest temperature increases are projected for the Iberian Peninsula, social vulnerability is likely to remain highest and adaptive capacity lowest (Fig. 5.24 and Table 6.2) in the countries of the southern Mediterranean and Middle East. Thus it is important that future research continues to focus on social dynamics and vulnerability, as well as on improvements in climate and biogeophysical modeling.

6.6 Research Needs and Gaps

The discussion of lessons learnt and key messages from the CIRCE case studies (Sect. 6.5) provides the basis for the identification of research needs and gaps. The major needs and gaps are outlined here, focusing on those that are relevant across all case-studies and the Mediterranean as a whole rather than those that are specific to individual case studies.

In particular, the CIRCE case-study work highlights the urgent need for improved observed data collection and monitoring. This encompasses the need for high-quality meteorological observations at high spatial and temporal resolutions (particularly for precipitation) and the same for marine observations (particularly relating to sea level, salinity, wave and storm regimes). There is also a need for improved environmental (biogeophysical) monitoring covering issues such as biodiversity, land use, water quality and saline intrusion. With respect to socio-economic information, the need is for more spatially disaggregated information (i.e., at finer than country resolution) and for longer data series. For some of these climate, biogeophysical and socio-economic data sets, it may be the case that data are being collected – here the issue is one of accessibility. Data may not be digitized, or are not deposited in freely/readily accessible archives. For all these types of data, there are also outstanding issues relating to data quality and homogenization. Assessing quality and implementing homogenization techniques is generally time consuming and any funding for such work tends to be rather limited. Yet good data are essential for reliable and robust integrated assessments

With respect to the CIRCE climate model output, this became available to the case-study teams at a fairly late stage of the project and time for analysis was limited. Thus more work on evaluating this output using observed (station and gridded) data sets is required. The work that was undertaken highlighted a number of biases in the models including a fairly systematic cold bias in temperature. Some of the case-study groups applied bias correction methods of varying complexity before using the model output, but future work would benefit from a more systematic and comprehensive approach using some of the more sophisticated bias correction methodologies that are being developed (Piani et al. 2010; Themeßl et al. 2011). In a number of cases, time prevented the use of all models for the indicators work and it would therefore be good to extend analyses to include all models. Future work with the CIRCE models would be facilitated by the availability of the output in a central archive with tools for extracting data subsets (i.e., particular variables and/or regions) before downloading. In addition, more work could be done comparing the fairly small CIRCE ensemble with larger RCM ensembles – in particular the ENSEMBLES RCM simulations (van der Linden and Mitchell 2009) and, when eventually available, CORDEX RCM simulations for the European and Mediterranean domains (Giorgi et al. 2009). Such comparisons would, *inter alia*, help to understand the impact of including air-sea coupling in the CIRCE regional simulations. The CORDEX RCM simulations and the CMIP5 global climate model simulations will allow extension of the CIRCE case-study assessments to consider the new Representative Concentration Pathways (Moss et al. 2010) which include mitigation scenarios and, potentially, to encompass fully initialized decadal predictions (Meehl et al. 2009). The availability of larger ensembles also opens up the possibility of constructing probabilistic projections and using other more sophisticated techniques for combining multi-model projections (Knutti et al. 2010).

In terms of both observed data and model projections, more work is clearly needed with respect to extreme weather and climate events and their impacts. Studying extreme events generally requires high temporal (typically daily) and spatial resolution data. This is particularly the case for events such as flash floods and urban flooding. More work is also needed on understanding urban heat island effects in Mediterranean cities.

Lack of time and the case-study teams' position as 'downstream users' mean that it has not been possible to fully exploit all outputs and products from CIRCE. Thus it would be useful to take full stock of CIRCE results such as those relating to water, tourism, agriculture and other relevant sectors. Similarly, future work with stakeholders could better exploit the more complex visualization, elicitation and decision-making tools developed within the CIRCE project.

Improvements in data availability and analysis as outlined above, should improve the potential for quantifying and refining the indicators identified for the case-study work – in particular those shown in the summary diagrams linking hazard, impact and vulnerability. In a number of cases, however, further work is still needed in order to disentangle (i.e., to understand and where possible quantify) climatic and non-climatic influences on the indicators. One major gap across the case studies is the lack of sub-national indicators encompassing vulnerability and adaptive capacity. In terms of working with indicators, a number of general research needs and gaps can be identified (see also Sects. 3.2.4 and 5.4) including a need for:

- Better definition and identification of thresholds and coping ranges
- More process-based rather than statistical modeling of indicators
- Where statistical modeling has to be used, more detailed consideration of non-stationarity, acclimatization and uncertainty issues
- More explicit consideration of indirect effects and feedbacks particularly when assessing complex cascading systems of impacts
- More explicit consideration of influences beyond the Mediterranean which may affect supply and demand within the region (e.g., energy, water, tourism, migration)

More detailed indicator information on vulnerability is required, particularly in urban Mediterranean environments, to plan for targeted adaptation at within-city scales. In particular, higher spatial resolution data is required to construct vulnerability indicators representing sub-populations. For example, health vulnerability could be stratified by age cohort (elderly, young) by income, education, and pre-existing illness. Vulnerability could also be stratified by physical location, e.g., coastal low-lying urban areas (prone to flood), communities with lower quality housing stock (greater risk of thermal stress; damage to extreme weather events), communities with limited access to resources (e.g., safe drinking water, health care services, air conditioning).

Adaptation is perhaps where the largest and most challenging research gaps and needs lie, including the need for:

- Development of cross-sectoral adaptation strategies and solutions, encompassing multiple temporal and spatial scales and cross-scale linkages
- More work on the implementation of adaptation strategies including barriers and enabling factors (taking account of decision-making processes and uncertainty)
- More quantitative and model-based evaluation of adaptation strategies (including costs)
- Consideration of linkages between adaptation and mitigation strategies and between these and sustainability strategies

There is scope to improve understanding of current coping and recovery strategies in the face of present-day climate hazards (especially extreme events at a regional level). This information could then be used as a basis to further develop best practice guidelines for regional adaptive responses to climate change. There also needs to be better understanding of how non-climate 'shocks' (such as the present political situation in some countries of North Africa and the Middle East, and economic difficulties such as those currently experienced in Greece) influence vulnerability and adaptive capacity.

Finally, future work on adaptation will benefit from a research and funding environment which facilitates more intensive and extensive collaboration and partnership working between the research and stakeholder communities.

References

Adger WN, Arnell NW, Tompkins E (2005) Successful adaptation to climate change across scales. Glob Environ Chang Part A 15(2):77–86. doi:10.1016/j.gloenvcha.2004.12

Afifi T (2009) Egypt case study: report for the environmental change and forced migration scenarios project. http://www.each-for.eu/

Biesbroek GR, Swart RJ, Carter TR, Cowan C, Henrichs T, Mela H, Morecroft MD, Rey D (2010) Europe adapts to climate change: comparing national adaptation strategies. Glob Environ Chang 20(3):440–450, Governance, Complexity and Resilience, doi:10.1016/j.gloenvcha.2010.03.005

Brooks N, Adger WN, Kelly PM (2005) The determinants of vulnerability and adaptive capacity at the national level and the implications for adaptation. Glob Environ Chang 15(2):151–163

Brown A, Gawith M, Lonsdale K, Pringle P (2011) Managing adaptation: linking theory and practice. UK Climate Impacts Programme, Oxford, UK, 47 pp. Available from: http://www.ukcip.org.uk/wordpress/wp-content/PDFs/UKCIP_Managing_adaptation.pdf

EEA (2010) 3 Adaptation. In: Isoard S (lead ed) The European environment state and outlook 2010: adapting to climate change. European Environment Agency, Copenhagen, Denmark, pp 22–44

Engle NL, Lemos MC (2010) Unpacking governance: building adaptive capacity to climate change of river basins in Brazil. Glob Environ Chang 20(1):4–13. doi:10.1016/j.gloenvcha.2009.07.001

Eriksen S, Aldunce P, Bahinipati CS, Martins RD'A, MolefeJI, Nhemachena C, O'brien K, Olorunfemi F, Park J, Sygna L, Ulsrud K (2011) When not every response to climate change is a good one: Identifying principles for sustainable adaptation. In: Climate and development, sustainable adaptation to climate change: prioritising social equity and environmental integrity, Earthscan, Abingdon, Oxforshire, UK/New York, NY, USA, pp 7–20, doi: 10.3763/cdev.2010.0060

European Commission (2007a) Communication from the Commission to the Council, the European Parliament, the European economic and social committee and the committee of the regions. Limiting global climate change to 2 degrees Celsius - The way ahead for 2020 and beyond. Commission of the European Communities, Brussels, Belgium. http://eur-lex.europa.eu/LexUriServ/LexUriServ.do?uri=CELEX:52007DC0002:EN:NOT

European Commission (2007b) Adapting to climate change in Europe – options for EU action {SEC(2007) 849}. Green paper from the Commission to the Council, the European Parliament, the European Economic and Social Committee and the Committee of the Regions. Commission of the European Communities, Brussels, Belgium. http://eur-lex.europa.eu/LexUriServ/LexUriServ.do?uri=CELEX:52007DC0354:EN:NOT

European Commission (2009) Commission staff working document accompanying the White paper – Adapting to climate change: towards a European framework for action – Impact assessment. Commission of the European Communities, Brussels, Belgium, 133 pp. http://ec.europa.eu/environment/climat/adaptation/index_en.htm

Füssel HM, Klein RJT (2006) Climate change vulnerability assessments: an evolution of conceptual thinking. Clim Chang 75(3):301–329

Giorgi F, Jones C, Asrar GR (2009) Addressing climate information needs at the regional level: the CORDEX framework. WMO Bull 58(3):175–183

Huq S, Rahman A, Konate M, Sokona Y, Reid H (2003) Mainstreaming adaptation to climate change in least developed countries (LDCs). International Institute for Environment and Development, London, p 40

Ionescu C, Klein RJT, Hinkel J, Kumar KSK, Klein R (2009) Towards a formal framework of vulnerability to climate change. Environ Model Assess 14(1):1–16. doi:10.1007/s10666-008-9179-x

IPCC (2007) Appendix I: Glossary. In: Parry ML, Canziani OF, Palutikof JP, van der Linden PJ, Hanson CE (eds) Climate change 2007: impacts, adaptation and vulnerability. Contribution of working group II to the fourth assessment report of the intergovernmental panel on climate change. Cambridge University Press, Cambridge, p 869

King R, Patterson G (1998) Diverse paths: the elderly British in Tuscany. Int J Popul Geogr 4:157–182

Klein R, Schipper L, Dessai S (2005) Integrating mitigation and adaptation into climate and development policy: three research questions. Environ Sci Policy 8:579–588

Knutti R, Abramowitz G, Collins M, Eyring V, Glecker PJ, Hewitson B, Mearns L (2010) Good practice guidance paper on assessing and combining multi model climate projections. In: Stocker TF et al (eds) Meeting report of the intergovernmental panel on climate change expert Meeting on assessing and combining multi-model climate projections. IPCC Working Group I Technical Support Unit, University of Bern, Bern

Lempert RJ, Groves DG, Popper SW, Bankes SC (2006) A general, analytic method for generating robust strategies and narrative scenarios. Manag Sci 52(4):514–528

McEvoy D, Lonsdale K, Piotr Matczak P (2008) Adaptation and mainstreaming of EU climate change policy: an actor-based perspective. CEPS policy brief, 149, January 2008, Centre for European Studies, Brussels, 15 pp. http://www.ceps.eu/system/files/book/1593.pdf

McLeman R, Smit B (2006) Migration as an adaptation to climate change. Clim Chang 76(1–2):31–53. doi:10.1007/s10584-005-9000-7

Meehl GA et al (2009) Decadal prediction: can it be skillful? Bull Am Meteorol Soc 90:1467–1485

Moser SC, Ekstrom J (2010) A framework to diagnose barriers to climate change adaptation. PNAS 107(51):22026–22031. doi:10.1073/pnas.1007887107

Moss RH, Edmonds JA, Manning MR, Rose SK, van Vuuren DP, Carter TR, Emori S, Kainuma M, Kram T, Meehl GA, Mitchell JFB, Nakicenovic N, Riahi K, Smith SJ, Stouffer RJ, Thomson AM, Weyant JP, Wilbanks TJ (2010) The next generation of scenarios for climate change research and assessment. Nature 463:747–756

Nelson V, Lamboll R, Arendse A (2008) Climate change adaptation, adaptive capacity and development discussion paper. DSA-DFID policy forum 2008. Available from: http://climateanddevelopment.nri.org/background_papers/nelson_lamboll_and_arendse_climate_change_adaptation.pdf

Piani C, Haerter JO, Coppola E (2010) Statistical bias correction for daily precipitation in regional climate models over Europe. Theor Appl Climatol 99:187–192

PNACC (2006) The Spanish national climate change adaptation plan. Oficina Española de Cambio Climático, Spain. http://www.marm.es/portal/secciones/cambio_climatico/areas_tematicas/impactos_cc/pnacc.htm

Ray P, Kirshen P, Vogel R (2010) Integrated optimization of a dual quality water and wastewater system. ASCE J Water Resour Plann Manag 136(1):37–47

Sgobbi A, Carraro C (2008) Climate change impacts and adaptation strategies in Italy: an economic assessment. FEEM Fondazione Eni Enrico Mattei research paper, CMCC research paper no 14, 1 Jan 2008

Stephens S, Ramsay D (2006) Predicting storm events at the coast. Water Atmos 14(4):14–15

Themeßl MK, Gobiet A, Leuprecht A (2011) Empirical-statistical downscaling and error correction of daily precipitation from regional climate models. Int J Climatol. doi:10.1002/joc.2168

Travers A, Elrick C, Kay R (2010) Position paper: climate change in coastal zones of the Mediterranean. Split, Priority Actions Programme, 2010, 72 pp. Accessed from: http://www.pap-thecoastcentre.org/

UNDP (2009) Human Development Report 2009, Overcoming barriers: Human mobility and development. United Nations Development Programme, New York, USA. 217 pp. Accessed from: http://hdr.undp.org/en/media/HDR_2009_EN_Complete.pdf

UNESCO (2009) National adult literacy rates published online: http://stats.uis.unesco.org/unesco/TableViewer/tableView.aspx

van der Linden P, Mitchell JFB (eds) (2009) ENSEMBLES: climate change and its im-pacts: summary of research and results from the ENSEMBLES project. Exeter EX1 3PB, Met Office Hadley Centre, UK, FitzRoy Road, 160 pp. Accessed from: http://ensembles-eu.metoffice.com/docs/Ensembles_final_report_Nov09.pdf

WHO-APAT (2007) In: Wolf T, Menne B (eds) Environment and health risks from climate change and variability in Italy. World Health Organization Regional Office for Europe, Copenhagen, Denmark, 106 pp. http://www.euro.who.int/document/E90707.pdf

Wilby RL, Dessai S (2010) Robust adaptation to climate change. Weather 65(7):180–185

Yamout G, El-Fadel M (2005) An optimization approach for multi-sectoral water supply management in the Greater Beirut Area. Water Resour Manag 19:791–812. doi:10.1007/s11269-005-3280-6

Chapter 7
Executive Summary

Clare M. Goodess, Maureen D. Agnew, Debbie Hemming, and Christos Giannakopoulos

Abstract Within the framework of the CIRCE project, 11 case-study locations were selected to reflect three generic environments (urban, rural and coastal), to quantify current and future climate change and to undertake an integrated assessment of the potential consequences to human communities and ecosystems at the regional to local scale. A rigorous common framework, referred to as the CIRCE Case studies Integrating Framework, was developed to facilitate a structured and systematic approach, based on indicators and the concept of vulnerability. Natural and human systems in all 11 case studies were found to be vulnerable to current climate variability and change as well as to social dynamics or drivers. In the absence of mitigation or adaptation, all case studies will experience continuing and increasing vulnerability to climate change.

Keywords Climate change • Mediterranean • Integrated assessment • Vulnerability • Adaptation

The heterogeneous nature of the Mediterranean environment, combined with a wide diversity of economic, social and cultural identities, make this region particularly amenable to integrated research on climate change impacts, vulnerabilities, and adaptive response. Within the framework of the CIRCE project, 11 case-study locations

---

C.M. Goodess (✉) • M.D. Agnew
Climatic Research Unit, School of Environmental Sciences,
University of East Anglia, UEA, Norwich, UK
e-mail: c.goodess@uea.ac.uk

D. Hemming
Climate Impacts Analysis Team, Met Office Hadley Centre, Exeter, Devon, UK

C. Giannakopoulos
Institute for Environmental Research and Sustainable Development,
National Observatory of Athens, Athens, Greece

were selected to reflect three generic environments (urban, rural and coastal), to quantify current and future climate change and to assess the potential consequences to human communities and ecosystems at the regional to local scale.

The case studies were chosen to reflect the east–west and north–south contrasts across the Mediterranean, using common selection criteria: vulnerability to climate change; availability of appropriate data; and access to regional stakeholders including decision and policy makers. The following case studies were selected:

- Urban case studies
    - Athens (Greece)
    - Beirut (Lebanon)
    - Alexandria (Egypt)

- Rural case studies
    - Tuscany (Italy)
    - Apulia (Italy)
    - Tel Hadya (Syria)
    - The Judean Foothills (Israel)

- Coastal case studies
    - Gulf of Valencia – Catalan Coast (Spain)
    - Gulf of Oran (Algeria)
    - Gulf of Gabès (Tunisia)
    - West Nile Delta (Egypt)

A rigorous common framework, referred to as the CIRCE Case studies Integrating Framework (CCIF) was developed to facilitate a structured and systematic basis for identifying and selecting indicators. Within this framework, climate dynamics is viewed as a key driver of changes in social and biogeophysical systems and is modulated by the inherent dynamics of these systems. The frequency and magnitude of climate hazards (sudden onset extreme events or longer-term changes) is altered by changes in the climate state and has measurable impacts on physical and social systems. Societal and environmental vulnerability to climate change is a function of the degree of exposure, the sensitivity of the system, and the capacity for adaptation.

The top-down, indicator-based approach was complemented by a bottom-up approach involving local and regional stakeholders. A participatory level of involvement was aimed for, with stakeholder dialogue on an informal basis throughout the project, culminating in a series of more formal regional stakeholder workshops. Stakeholders contributed to a number of areas of work, including: identification and selection of indicators and their linkages; provision of data and knowledge; identification of indicator thresholds; and, identification and assessment of adaptation strategies and policy recommendations. A series of briefing notes and information sheets suitable for stakeholders was produced during the project, and where necessary material was translated into the appropriate local language. Based on the CIRCE experience, a good practice checklist for facilitating stakeholder involvement in research projects has been produced.

7 Executive Summary

Identification and construction of physical and socio-economic indicators was the most challenging and time-consuming aspect of the case-study work. A detailed set of selection criteria was defined and the process of reviewing and refining indicators was iterative. Nonetheless, a number of data and methodological challenges were encountered. Availability of appropriate data of sufficient quality and length was an issue in many cases. Other challenges included combining and using indicators derived from both quantitative and qualitative data, and distinguishing between impact and vulnerability indicators. Despite these issues, indicator linkages diagrams provided a useful preparatory stage for structuring the integrated assessment for each case study. They also provided a useful visual communication tool for discussing issues of vulnerability and adaptation with regional stakeholders and for the identification of areas of uncertainty and knowledge gaps.

The selected indicators were classified as follows:

- Climate and atmosphere indicators
  - Core climate indicators: maximum temperature and precipitation (for a key season relevant to the case study)
  - Additional case-study specific climate indicators (e.g., hot days, hot nights, drought index, greatest 3-day rainfall)
- Biogeophysical indicators grouped under four themes:
  - Marine and coastal systems (e.g., sea surface temperature, alien marine species)
  - Terrestrial ecosystems and biodiversity (e.g., number of wildfires and area burnt, vegetation types)
  - Freshwater systems (e.g., water availability)
  - Agricultural and forestry systems (e.g., yields of key crops)
- Social indicators grouped under two themes:
  - Human health (e.g., all-cause mortality)
  - Economic systems including the tourism sector (e.g., a daily climate tourism potential index, socio-economic tourism losses), the energy sector (e.g., energy consumption by type and sector), fisheries and aquaculture (e.g., fish catch/landings), agricultural production (e.g., olive oil and wine), and the water industry (e.g., water consumption by sector).

In the first and major assessment stage, impacts and vulnerability due to exposure to hazards associated with current and recent climate variability and change were explored using observed data. Linkages and sensitivities between the specific climate impacts of relevance for each case-study type (urban, rural and coastal) were assessed and the current climate hazards and biogeophysical and social vulnerabilities representing the underlying drivers and site conditions identified. For some impacts, there are clear, direct links with climate events while for others the causal relationships are more indirect, via a cascade of climate, social and economic influences. Water availability and extreme temperatures are common drivers of current climate impacts across all case studies, including, for example, freshwater supply and heat stress for urban populations; irrigation capacity and growing season length for agricultural regions; and saltwater intrusion of aquifers and tourist visitor numbers at coastal locations.

The climate drivers were shown to manifest in a wide range of impacts depending on local characteristics and vulnerabilities. For some case-studies, specific impacts, hazards and/or vulnerabilities are observed, such as peri-urban fires in Greater Athens, infrastructure vulnerability to coastal flooding in Alexandria, groundwater levels in Tel Hadya and vector-borne diseases in the Gulf of Oran, while others are common across many of the case studies.

For the Athens urban case study, the energy industry is particularly susceptible to changes in the availability and variability of freshwater, whereas in the densely populated urban area of Beirut, water resources are largely utilized for human consumption and water quality is currently at risk by salinization of aquifers. Salinization is also a problem in Alexandria where fresh water resources are heavily competed for among domestic, industrial and tourist functions.

In rural locations, changes in the availability and quality of freshwater resources, together with other climate variables, were shown to exert key impacts on the rural economy via agricultural production of wheat (all four rural case study locations), barley (Tel Hadya), heavily irrigated vegetables including tomatoes (Apulia), and olives and grapes (Apulia and Tuscany).

For the coastal case studies, the impacts of coastal erosion, flooding and salinization, together with those on coastal ecosystems and fisheries are also of key importance to local communities, economies and ecosystems. For example, for the Gulf of Oran region in Algeria reductions in rainfall in the Chellif River basin since the 1970s are leading to concerns on how the changes in outflow are influencing the salinity, dynamics and ecosystems within the region. For the Gulf of Valencia Catalan coast in Spain, observed reductions in river flow during the latter half of the twentieth century, linked to abstraction for agricultural and domestic purposes, and concomitant decreases in river sediment transportation to the Ebro delta are associated with increased rates of coastal erosion and subsidence, further increasing the vulnerability of this delta region to sea level rise and salinization. In the Gulf of Gabès, invasion of alien jellyfish species pose a threat to numerous native Mediterranean species in the region. However, other invasive species, such as Spanish mackerel, have become economically important and helped to reduce the economic vulnerability of the area.

The case-study work on vulnerability to current and recent climate variability and change provided the context for the construction of climate-model based indicators for the future (to 2050 and for the A1B emissions scenario). First, climate indicators were constructed using output from the CIRCE global and regional climate model simulations for the Mediterranean. These have the main novel characteristic of incorporating a realistic representation of the Mediterranean Sea including coupling between the sea and atmosphere. The projections indicate warming, including more frequent hot extremes (very hot days and nights, heat waves) and fewer cold extremes (very cold days and nights, cold spells), across all the case studies. Warming is greatest in summer and over the Iberian Peninsula.

The projections for precipitation are less consistent and robust, although mean total precipitation either decreases or shows little change for the case studies. It is rather difficult to draw clear messages about changes in even the direction of precipitation extremes, though there is, for example, some indication of an increase in 3-day rainfall for Athens. For the coastal case studies, the projections indicate a warming of sea surface temperature and an increase in sea surface height due to this warming. There is no evidence, however, of future increases in marine storminess.

Using the CIRCE climate model output, together with regional climate model simulations from the earlier EU ENSEMBLES project, a number of the selected biogeophysical and social indicators were quantified for the future. For the Athens urban case study, for example, these indicate: increased risk of peri-urban fires; more ozone exceedance days; an increase in heat-related mortality; and, an increase in summer energy demand. For the Gulf of Gabès coastal case study, the indicators show: an increase in the number of alien marine species; a lengthening of the summer tourist season but with fewer 'favorable' days in summer (offset by a small increase in the number of 'highly favorable' days in winter); and, a consequent economic loss due to a reduction in tourism-related employment.

The climate projections which underlie the case-study impact and vulnerability assessments are inevitably subject to uncertainties. Using multiple models is important in this context, but the CIRCE models are subject to biases. In comparison to present-day observations, for example, the models have a systematic cold bias of about 2°C. Where possible, bias correction and other techniques have been used to allow for such biases. There are also uncertainties 'downstream' of the climate models and projections – largely relating to the modeling of impacts. In addition, there are uncertainties associated with all socio-economic projections used in the case studies – such as population projections. Additional uncertainties are introduced by the uncertain political and economic situations in some of the case studies at the time of writing. Thus there are uncertainties inherent to all stages of the integrated assessments and it is important to consider all these aspects in the context of adaptation decision making.

The case-study integrated assessments provided the basis for collaborative work with stakeholders focusing on adaptation towards the end of the project. Workshops, seminars and discussion groups were organized in the 11 case-study locations in local languages (or local languages and English). Background material on the case-study work (including briefing notes and information sheets) was distributed in advance or during the workshops. A variety of workshop approaches and tools were used, including plenary sessions, focus groups, expert interviews, 'world café' conversation methods and perception questionnaires.

A recurring theme among the workshop participants was the need for targeted regional information over both the short term (to better cope with current climate hazards) and the longer term (to guide adaptation planning for future climate hazards). An emphasis was also placed on the need for concerted adaptation actions at multiple spatial scales (local, regional and national) and across multiple sectors (e.g., water resources, agriculture, tourism), in addition to complementary mitigation measures (such as the use of renewable energy). Thus actions directed at reducing local social stressors (e.g., overfishing and marine pollution) could reduce wider vulnerability to current climate hazards and future climate change. Raising awareness and improving management were viewed as important facilitators in the implementation of adaptation plans. In this context, the workshops themselves contributed to building regional-level capacity in terms of providing and exchanging information and knowledge.

Formulating appropriate adaptive responses in the case studies was built on an understanding of current climate hazards, impacts and vulnerabilities specific to the region using the indicator-based approach to assess for climate sensitivity, trends and system thresholds. Thus, utilizing the evidence-based integrated assessment framework and in collaboration with regional stakeholders, a set of potential specific case-study adaptation options was identified, and a range of adaptation measures proposed across several sectors in the context of national adaptation strategies and the wider CIRCE Mediterranean work.

In line with the EU White Paper on adaptation, 'smart', 'green', and 'grey' adaptation options were identified at a case-study level. The outcomes of several of the stakeholder workshops (including the Gulf of Gabès, Alexandria and the West Nile Delta) support a coordinated 'smart' (behavioral-managerial-policy) adaptive response among government departments and organizations. These include plans to implement an integrated coastal zone management plan to protect low-lying areas of the West Nile Delta, and proposals to create an adaptation strategy framework across multiple spatial scales to complement 'green' adaptation initiatives (such as use of renewable energy, reducing overfishing and controlling the spread of invasive marine species) in the Gulf of Gabès. The Tel Hadya rural case study assessed the use of supplemental irrigation for rain-fed agricultural systems as a 'smart' adaptive option that utilizes knowledge of varying water requirements during specific stages of the crop cycle. 'Grey' technological based adaptive responses include coastal engineering structures to protect the Valencia-Catalan coast, and the construction of desalinization plants in the Judean Foothills and Greater Beirut Area to supplement freshwater resources. Ecosystem-based 'green' responses represent a broader category of adaptation actions that cover improved building efficiency in urban areas (Athens), widening sustainability practices among rural tourism-farming communities (Tuscany, Apulia), the use of drought-resistant or less water intensive crop varieties (Tuscany, the Judean Foothills), and actions to increase water availability, such as rainfall harvesting and wastewater reclamation in Tel Hadya and the Judean Foothills.

Two key messages with respect to adaptation emerged:

- There should be concerted adaptation strategies implemented at all levels from individual/community, to regional, national and transnational levels.
- Linking adaptation action into sustainability initiatives is a potentially useful means of reducing climate vulnerability in a preliminary stage.

Natural and human systems in all 11 case studies were found to be vulnerable to current climate variability and change as well as to social dynamics or drivers. The climate projections of increases in mean and extreme high temperature and decreases in precipitation are considered to be robust, although there is uncertainty with regards to the magnitude of change. They indicate that all case studies will experience continuing and increasing vulnerability to climate change in the absence of mitigation or adaptation. Projections for other extreme weather events, such as heavy precipitation and flooding, are highly uncertain, but any increase in such events would further increase vulnerability. Very few examples have been identified of where climate change might be beneficial for the case studies or where there might be opposing or compensating changes. At the same time, social dynamics and drivers such as population growth (at least in the short term and in the southern Mediterranean) are likely to further increase vulnerability. The local interplay and balance of factors will reflect the spatial complexities of the Mediterranean natural and human environments. In terms of broad geographical patterns, social vulnerability is likely to remain highest and adaptive capacity lowest in the countries of the southern Mediterranean and Middle East. Thus it is important that future research continues to focus on social dynamics and vulnerability, as well as on improvements in climate and biogeophysical modeling.

The lessons learnt and key messages from the CIRCE case studies provide the basis for the identification of research needs and gaps. In particular, the CIRCE case-study work highlights the urgent needs for improved observed data collection and monitoring. Other needs relate to the analysis and availability of climate projections for the region, and to the quantification and interpretation of biogeophysical and vulnerability indicators. Adaptation is, however, where perhaps the largest and most challenging research gaps and needs lie. This work will benefit from a research and funding environment which facilitates more intensive and extensive collaboration and partnership working between the research and stakeholder communities.

Appendices

Appendix 1: The CIRCE Case-Study Abbreviations

Urban case studies:

AT   Athens, Greece
AL   Alexandria, Egypt
BE   Beirut, Lebanon

Rural case studies:

TU   Tuscany, Italy
AP   Apulia, Italy
TH   Tel Hadya, Syria
JF   Judean Foothills, Israel

Coastal case studies:

VA   Gulf of Valencia – Catalan coast, Spain
OR   Gulf of Oran, Algeria
GA   Gulf of Gabès, Tunisia
WN   West Nile Delta, Egypt

Appendix 2: Key Case-Study Indicators

Table A2.1 Climate indicators

Indicator name	Case study	Indicator description[a]
Maximum temperature (Tn)	AT	Annual, 1897–2006; Seasonal, 1897–2007; National Observatory of Athens
	BE	Annual, seasonal; 1950–2001, Aéroport International de Beyrouth
	AL/WN	Monthly; 1973–2008, El Nozhah
	TU	Seasonal, 1951–2006, E-OBS gridded data (25 km)
	AP	Monthly; 1951–2005, 11 stations in Apulia
	JF	Annual, seasonal; 1951–2008; 3 sites: north (Beit-Jamal), central (Beit-Guvrin) and south (Lahav)
	TH	Wheat growing season (Oct–Jun), 1978–2007, Tel Hadya
	VA	Annual, monthly; 1950–2008 (Ebro; Fabra)
	OR	Annual, 1926–2007, Oran
	GA	Seasonal, monthly; 1973–2008 (Sfax)
Minimum temperature (Tx)	AL/WN	Monthly, 1973–2008, El Nozhah
	TU	Seasonal, 1951–2006, E-OBS gridded data (25 km)
	AP	Monthly; 1951–2005, 11 stations in Apulia
	JF	Annual, seasonal; 1951–2008; 3 sites: north (Beit-Jamal), central (Beit-Guvrin) and south (Lahav)
	TH	Wheat growing season (Oct–Jun), 1978–2007, Tel Hadya
	VA	Annual, monthly; 1950–2008 (Ebro; Fabra)
Mean temperature	AL	Monthly, 1973–2008, El Nozhah
	TU	Seasonal, 1951–2006, E-OBS gridded data (25 km)
	TH	Wheat growing season (Oct–Jun), 1978–2007, Tel Hadya
	VA	Annual, monthly; 1950–2008 (Ebro; Fabra)
	GA	Annual, seasonal; 1950–2008 (Gabès)
Precipitation	AT	Annual; 1899–2007, National Observatory of Athens
	BE	Annual, 1950–2001, Aéroport International de Beyrouth
	AL/WN	Annual, 1973–2008, El Nozhah
	TU	Seasonal; 1951–2006, E-OBS gridded data (25 km)

(continued)

Table A2.1 (continued)

Indicator name	Case study	Indicator description[a]
	AP	Seasonal, annual; 1951–2005, 24 stations in Apulia
	JF	Annual, 1951–2008, 3 sites: north (Beit-Jamal), central (Beit-Guvrin) and south (Lahav)
	TH	Wheat growing season; vegetative growth season; reproductive season; 1978–2007, Tel Hadya
	VA	Annual, seasonal; Fabra 1914–2008, Ebro 1906–2008
	OR	Annual, 1926–2007; Oran
	GA	Seasonal; 1973–2008 (Sfax; Gabes)
Probability of rainfall occurrence	OR	Daily, 1941–1970; 1971–2000; Oran
Standardized Precipitation Index (SPI)	TH	Wheat growing season (Oct–Jun); reproductive period (Feb–Apr); 1979–2007, Tel Hadya
Greatest 3-day rainfall	AT	Annual, 1899–2007, National Observatory of Athens
	BE	Annual, 1950–2001, Aéroport International de Beyrouth
Longest dry spell	TU	Maximum number of consecutive days with precipitation amount <1 mm, 1951–2006; E-OBS gridded data (25 km), Tuscany region
Very heavy rainfall days (>95th percentile)	AT	Annual, 1899–2007; National Observatory of Athens
Hot (very) days/nights	AT	Very hot days (>37°C), Hot days (>35°C), Very hot nights (>25°C), Hot nights (>20°C); 1897–2007
	BE	Hot days (>31.4°C); Hot nights (>25.4°C); 1950–2001; Aéroport International de Beyrouth
	AL/WN	Hot days; 1973–2008, El Nozhah
	TU	Hot days, hot nights, E-OBS gridded data (25 km)
	AP	Very hot days; E-OBS data, 1951–2005
	JF	Very hot days, very hot nights, 3 sites: north (Beit-Jamal), central (Beit-Guvrin) and south (Lahav)
	OR	Hot days; 1926–2007; Oran
	GA	Very hot days (24-h; 5 % upper limit: ~34–36°C)
Very cold nights	AP	Consecutive days Tn<5th percentile; E-OBS data, 1951–2005
Cold Weather Duration Index (CWDI)	AP	Days Tn<5th percentile; E-OBS data, 1951–2005

(continued)

Table A2.1 (continued)

Indicator name	Case study	Indicator description[a]
Urban Heat Island (modelled urban/rural differences)	AT	Hot nights (Tn >25°C), Hot days (Tx >40°C); Met Office urban surface scheme coupled to a RCM
	BE	Hot nights (Tn >25°C), Hot days (Tx >40°C); Met Office urban surface scheme coupled to a RCM
	AL	Hot nights (Tn >25°C), Hot days (Tx >40°C); Met Office urban surface scheme coupled to a RCM
Heat Wave Duration Index (HWDI)	TU	Maximum annual number of consecutive days >30°C (95th percentile); E-OBS gridded data (25 km)
	AP	Consecutive days Tx > 95th percentile; E-OBS data 1951–2005
Tourist season: Seasonal Shift Index	GA	Dates of the beginning (23°C) and end (<23°C) of the tourist season, Sfax, Gabès, Djerba
Tourist season: Seasonal Shift Index	AP	Dates of the beginning (23°C) and end (<23°C) of the tourist season, using Tmean for Apulia region
Storm frequency (lightening and thunder reported)	OR	Annual, seasonal, 1950–2007; Oran

[a]Details are given for observed data used for constructing indicators

Table A2.2 Marine indicators

Indicator name	Case study	Indicator description[a]
Relative sea level	VA	Storm surge – hindcast annual mean sea level (cm), 1958–2001, three sites
	GA	Monthly; 1999–2007; Sfax Harbour
Significant wave height (Hs)	VA	Wave storms (high Hs 1.5 m; Very high Hs 2.0 m) – number and duration for 2 sites: Cap Tortosa, Ebro Delta; 1990–2006
Average wave direction	VA	Llobregat, Cap Tortosa, Cullera Bay; frequency of occurrence
Maximum significant swell height	GA	Annual (3 sites in the Gulf of Gabès); 1974–1994
Sea water temperature	GA	Annual (depths <30 m), 1920–2007, all available sites (from MEDATLAS and ISTM)
Sea surface temperature (SST)	VA	Annual mean, min, max SST for l'Estartit 1969–2008 (1972/1973 missing). Annual mean SST for Cap Tortosa 1990–2008

[a]Details are given for observed data used for constructing indicators

Table A2.3 Coastal impact and vulnerability indicators

Indicator name	Case study	Indicator description[a]
Marine water quality	VA	Dissolved oxygen depletion, chlorophyll a (eutrophication), CO_2 (pH) in coastal/estuarine waters
Coastal waste water discharge	OR	Discharge into the Bay of Oran (m^3/day) by source (domestic; industrial) for the years 1995 and 2005
Alien marine species	GA	Number of species per year of observation, Gabès Gulf; 1966–2007 (adjusted for reporting bias)
Invasive marine species	OR	Jelly fish; green algae (plant survey 2006–2010; discussion with fisherman and wider Mediterranean studies)
Food-web alterations	OR	Three previous studies describing the gut contents (% different prey species) of Odontocetes cetaceans (common and bottlenose dolphins)
Coastal Vulnerability Index	GA	Five classes from 1 (low) to 5 (very high); 5 coastal segments of the Gabès Gulf
Shoreline modification	GA	Aircraft photographs of shoreline for at least two reference years (>40y separation); 3 sites
Coastal erosion	VA	Length of dynamic coastline 2004; Spatial variation in erosion state along the Catalan coast; DEDUCE project, EUROSION database
Vulnerable littoral sites/ buffer zones	OR	La Macta Nature Reserve – ecological/ water quality monitoring
Vulnerable littoral sites/ buffer zones	OR	The Cheliff River (longest river in Algeria)/Estuary: identification of pollution sources; water quality monitoring (21 parameters), 2004–2005
Saline intrusion	BE	Monthly number (and %) of wells sampled (20–22) with Total Dissolved Solids (TDS) at low, medium and high salinity levels, Oct 2004–Dec 2005; Greater Beirut Area
Salt-water intrusion	VA	Percentage of salt-wedge intrusion time in the Ebro River; 01-01-99 to 29-02-00
Coastal zone land use	VA	Extent of the Maritime Terrestrial Public Domain, 2004 onwards
Land vulnerable to flooding	WN	Percentage of land cover (six classes) vulnerable to sea-level rise, West Nile Delta

[a] Details are given for observed data used for constructing indicators

Table A2.4 Terrestrial ecosystem impact and vulnerability indicators

Indicator name	Case study	Indicator description[a]
Water availability	TU	Annual, estimated using annual evapotranspiration and total rainfall, sub-regional; 1999, 2003, 2008, Tuscany region
Groundwater levels	TH	Water table level (m below the ground); annual; 1983–2007
Environmental water budget	AP	Total precipitation minus evapotranspiration (derived from temperature, calibrated for Apulia); Annual; 1950–2000
Runoff	OR	Annual runoff, 1950–2004, hydrological station at Sidi Ali Benyoub (close to Oran)
Fire Weather Index (FWI)	AT	Daily (based on Canadian Fire Weather Index), 1899 onwards, Athens
Fire Weather Index (FWI)	TU	Annual, 1984–2005
Fire frequency	AT	Daily, 1983–1990; as a function of FWI; Attica region
Fire frequency	TU	Annual number of wildfires in Tuscany, 1984–2007
Fire surface area burnt	AT	Annual, 1955–2007; Greece
Fire surface area burnt	TU	Size of area burnt (ha) by Tuscan administrative municipality; annual 1984–2007
Wooded area burnt	AP	Size area burnt (ha); annual; 2006–2009; Apulia
Fire occurrence	AL/WN	MODIS fire pixel count, monthly, 2001–2004, Greater Nile Delta region
Ozone exceedance	AT	Daily 8-h max concentration, 1990–2000; annual exceedance days (>61 ppb)
Ozone levels	AL/WN	Total ozone (DU) column anomaly, monthly, 1997–2007
Aerosol characterisation	AL	Aerosol type (dust, pollution, mixed, clear), for selected dates 2000–2003
Atmospheric loading; aerosol optical depth (AOD)	AL	Daily, 2000–2009, MODIS satellite data (Level 2) over Alexandria
Land use	TU	CORINE land cover classes, level 1, 2006
Land use	JF	Land cover (%, 8 classes), 2002
Land use	AP	Land cover (%, 8 classes), 2005
Land use	OR	Land cover (%, 7 classes) in the Bay of Oran
Forest tree growth (stem volume of Jerusalem Pine)	JF	13 measurements from selected plantations along a rainfall gradient, Judean Foothills
Tree health	JF	Case study 1999; survey of forest plantations

(continued)

Table A2.4 (continued)

Indicator name	Case study	Indicator description[a]
Vegetation types in natural ecosystems	JF	Four vegetation surveys along a rainfall gradient (six-fold classification of vegetation – presence or absence – for mean annual rainfall 300–481 mm)
Carbon sequestration	JF	Organic carbon stock (kg m^{-2}) in topsoil (10 cm depth) in 2 dwarf shrublands, 2 sites (north, south)
Carbon uptake	OR	Monthly mean carbon uptake by vegetation type (g)

[a]Details are given for observed data used for constructing indicators

Table A2.5 Agricultural impact and vulnerability indicators

Indicator name	Case study	Indicator description[a]
Rain-fed durum wheat yield	TU	Annual, 1996–2008; Tuscany region
Rain-fed wheat grain yield	JF	Annual, 2–7 years of data for different varieties of wheat; northern region and southern region
Rain-fed wheat grain yield	TH	Annual (mean tonnes/ha); survey data for Tel Hadya, 1994/1995–2008/2009
Above ground biomass development	TH	Growing season 1984–1985, Tel Hadya
Wheat production	AP	Annual (tonnes); Apulia, 1980–2000
Wine (total) production	AP	Annual (hectolitre); Apulia, 1980–2000
Olive oil production	AP	Annual (kg); Apulia, 1980–2000
Grain yield	OR	Annual, 1984–2004, Sidi Belabbès plain (close to Oran)

[a]Details are given for observed data used for constructing indicators

Table A2.6 Human health impact and vulnerability indicators

Indicator name	Case study	Indicator description[a]
All-cause mortality	AT	Daily, 1992–2006; Athens district
Life expectancy at birth	AT	Five-year bands 1950–2005, Greece
All-cause mortality	BE	1997–1999; Greater Beirut
Water-borne disease	BE	Monthly; total cases of water-borne disease; 2001–2010; Lebanon
Food-borne disease	BE	Monthly; cases (dysentery, typhoid, hepatitis A) of food-borne disease; 2001–2010; Lebanon
Total cost of water-borne disease	BE	Annual cost of morbidity (US$), extrapolated from the years 2000 to 2030
Mortality maximum temperature thresholds	VA	Spatial variation (seven sub-regions) in maximum temperature thresholds related to mortality
All-cause mortality	OR	All-cause mortality in summer for Oran city, 1982–2010
Life expectancy at birth	GA	Five-year bands 1950–2005, Tunisia

[a]Details are given for observed data used for constructing indicators

Table A2.7 Demographic vulnerability indicators

Indicator name	Case study	Indicator description[a]
Urban population growth	AT	Annual; national, 1950–2005 (5-year periods)
Population growth/density/settlement size	JF	Annual; 1948–2008 for the Judean Foothills
Population growth	VA	Annual; five large urban areas within the Catalan coastal zone
Population growth	AP	Decadal; 1921–2008; Apulia
Population	OR	Annual population of Oran city, 1950–2000
Population growth	GA	Annual population growth rate, 1950–2000, Tunisia
Urbanisation	GA	Annual urban/rural population, 1950–2000, Tunisia

[a]Details are given for observed data used for constructing indicators

Table A2.8 Socio-economic impact and vulnerability indicators

Sector; Indicator name	Case study	Indicator description[a]
Tourism		
Daily Climate Tourism Potential Index (DCTPI)	GA	Annual % by class (highly favourable; favourable; acceptable, etc.) for Djerba, 1973–2003
Socio-economic losses due to changes in DCTPI	GA	Theoretical simulation of change in loss of capital; loss of annual gains; reduction in employment (between 1960 and 2007); Djerba-Zarzis area
Socio-economic losses due to beach erosion	GA	Theoretical simulation of change in loss of capital; loss of annual gains; reduction in employment (between 1960 and 2007); Djerba-Zarzis area
Tourist arrivals (domestic; international)	TU	Annual number of tourist stays in rural accommodation in Tuscany, 1993–2007
Tourist arrivals	AP	Annual; 1998–2007; Apulia
Tourist visits (domestic; international)	JF	Tourist visits to Bet Guvrin national park, annual, 2003–2008
Energy		
Energy consumption (domestic and commercial)	AT	Daily, 1993–2001; Greater Athens Area (Attica)
Electricity generation and consumption	VA	Annual, 1980–2009, Spain

(continued)

Table A2.8 (continued)

Sector; Indicator name	Case study	Indicator description[a]
Water		
Water resources by type	JF	Includes lake, wells, other surface, reclaimed waste water 1990–2007 (annual from 2003, irregular intervals prior to 2003);
Water consumption by sector	JF	Sectors include: agricultural, domestic and public, industrial; annual from 2003 (irregular intervals prior to 2003); 1964/1965–2007
Water consumption by sector	TU	Sectors: agriculture; civil; industry; annual; Agriculture: 2000–2008; Industry: 1995–2007; Civil: 2006
Water consumption	AP	Water consumption per capita, Annual; 2000–2008; Apulia (main towns)
Water rationing	AP	Annual; Percentage of main towns in Apulia rationing water, 2000–2008
Insurance		
Insurance claims	VA	Insurance claims/charges for property damage, 1971–2009, Spain
Fishing industry/aquaculture		
Fish production	GA	Fish catch (tons/year) by species (Caramote prawn, grey triggerfish, hake, common Pandora) 1995–2006; Gulf of Gabès
The Economy		
Gross Domestic Product (GDP)	AT	Annual GDP in billions/GDP per capita, for Greece, 2008–2010
Gross Domestic Product (GDP)	TU	Sector composition (%) of GDP for Tuscany, the year 2004
Gross Domestic Product (GDP)	JF	Annual GDP constant prices/current prices, for Israel, 1980–2010
Gross Domestic Product (GDP)	GA	Annual GDP based on purchasing power parity for Tunisia, 2008–2010
Employment	JF	Annual number of persons, for Israel, 1980–2010
Unemployment rate	JF	Annual %, for Israel, 1980–2010

[a]Details are given for observed data used for constructing indicators

Index

A
Adaptation
    lessons learnt, CIRCE case studies, 192–197
    linking impacts and vulnerability
        and adaptive capacity, 167–170
        indicators, coping range and, 166–167
    local and regional scales
        case-study information consolidation, 181–187
        CIRCE stakeholder perspective, 171–181
    outside case-study regions
        global effects, regional tourism, 187–188
        governance issues, 188–189
        migration, 188
        policy for, 189–192
        supply and demand issues, 187
    research needs and gaps, 197–199
    stakeholder involvement, 33–34
Air-pollution, 71, 133–135
Alexandria (Egypt), 13
Apulia
    location, 15
    wine production, 77
Athens urban case study
    biogeophysical and social vulnerability indicators, 68
    CIRCE case studies
        air-pollution, 133–135
        energy demand, 138–140
        human health risk, 135–138
        peri-urban fire, 131–133
    daily energy consumption and daily mean air temperature, 67
    location, 12
    Mediterranean environment, location, 12
    summary of, 206
    summer, climate projections, 114–116

B
Beirut
    heat stress and water pollution, 69
    location, 12–13
Biogeophysical impact
    capacity, 124–126
    challenges, 123
    changes
        for future, 144–145
        social systems and communities, 145–148
    dynamics, 9
    indicators
        changes, 54
        climate signal isolation, 55
        coastal case-study, 54
        data availability, 54
        organization, 42
        rural case-study, 53–54
        urban case-study, 53
    integrated assessment, 126
    Mediterranean changes, 127–131
    quantitative, 123
    vulnerabilities, climate impact
        coastal zones, 97–101
        rural region, 80–88
        urban region, 70–73

C
Carbon sequestration, Judean foothills, 76

CCIF. *See* CIRCE Case studies Integrating Framework (CCIF)
CIRCE case studies
  Athens urban case study
    air-pollution, 133–135
    energy demand, 138–140
    human health risk, 135–138
    peri-urban fire, 131–133
  biogeophysical and socioeconomic projections
    Athens urban case study, 131–140
    Gulf of Gabès coastal case study, 140–143
  coastal zone (*see* Coastal zones, climate impact)
  Gulf of Gabès coastal case study
    socio-economic consequences, 142
    tourism, 141–143
  project summary of, 203–209
  rural zone (*see* Rural zone, climate impact)
  stakeholder involvement, 27
  urban zone (*see* Urban region, climate impact assessments)
CIRCE Case studies Integrating Framework (CCIF), 8–9, 204
Climate and atmosphere indicators, 51–52. *See also* Indicators
Climate change. *See* Indicators
Climate dynamics, 9
Climate hazards
  coastal zones
    climate related impacts, 90
    DCTPI, 97
    sea level indicator, 92, 93
    sea surface temperature, 92
    thermal comfort, 97
  definition, 9
  rural zone
    CIRCE rural case study areas, 79
    precipitation and evaporation, 78–79
    rainfall, 80
    temperature, 79–81
  urban region, 69–70
Climate impact assessments
  coastal zones
    biogeophysical and social vulnerabilities, 97–101
    climate hazards, 91–97
    climate related impacts, 88–91
  future projections
    assessment uncertainties, 151–160
    biogeophysical and socioeconomic, 123–131

CIRCE case studies, 131–143
climate projections, 107–123
key projected changes, 143–145
linking vulnerabilities and impacts, 148–150
social systems and communities, 145–148
  rural region, 73–75
    biogeophysical and social vulnerabilities, 80–88
    climate hazards, 78–80
    climate related impacts, 75–78
  urban region
    biogeophysical and social vulnerabilities, 70–73
    climate hazards, 69–70
    climate related impacts, 65–69
Climate projections
  case-study impact, 207
  CIRCE models, 108–109
  Global Climate Model (GCM), 107
  Mediterranean-wide analysis, 112
  Palmer Drought Severity Index (PDSI), 119–120
  precipitation level, 109–110
  sea surface height (SSH), 122
  sea surface temperature (SST), 120–121
  summer, in Athens, 114–116
  temperature level, 109–113
  winter season, 116
Coastal zones, climate impact
  biogeophysical and social vulnerabilities
    CIRCE coastal case study locations, 99–100
    jelly fish, 101
    low-lying lands, 98–100
  case study summary, 206
  climate hazards
    DCTPI, 97
    sea level indicator, 92, 93
    sea surface temperature, 92
    thermal comfort, 97
  climate related impacts
    economic impacts, 91
    hazards and vulnerability, 90
    waste-water discharge, 91, 92
  Gulf of Gabès, Tunisia, 18
  Gulf of Oran, Algeria, 18, 95–97
  Gulf of Valencia-Catalan coast, Spain, 17
  population density, 89
  West Nile Delta, Egypt, 19

Index

D
Daily Climate-Tourism Potential Index (DCTPI), 97

E
Economic impacts. *See* Social vulnerabilities, climate impact
Ecosystem, climate related impacts, 75

F
Fire Weather Index (FWI), 132

G
Gulf of Gabès coastal case study
  coastal zones, climate impact, 18
  location, 18
  socio-economic consequences, 142
  tourism, 141–143
Gulf of Oran, Algeria, 18, 95–97
Gulf of Valencia-Catalan coast, Spain, 17

H
Hazards. *See* Climate hazards

I
Indicators
  biogeophysical (*see also* Biogeophysical impact)
    challenges, 54–55
    summary of, 205
    themes, 52–54
  CIRCE case-study, organization of, 42
  climate and atmosphere, 51–52
  identification and construction of, 205
  methodology
    coping mechanisms, 45
    data and methodological challenges, 45–47
    integrated vulnerability assessment (*see* integrated vulnerability assessment)
    methods of presentation, 47–48
    reviewing and refining, 43
    selection criteria, 42–43
    threshold, 44–45
    trends, 43–44
  social
    challenges, 57–59
    summary of, 205
    themes, 55–57
  summary of, 205
Integrated vulnerability assessment. *See also* Climate impact assessments
  interactive stakeholder approach, 49
  linkages diagram, 49–50
  potential indicators, 48

J
Judean foothills
  carbon sequestration, 76
  location, 16
  Mediterranean environment, case study, 16

L
Linking impacts and vulnerability, adaptation
  and adaptive capacity, 167–170
  thresholds, coping range and, 166–167

M
Mediterranean environment, case study. *See also* CIRCE case studies
  agriculture, forest and ecosystem services, 127–128
  CCIF, 8–9
  coastal region
    Gulf of Gabès, Tunisia, 18
    Gulf of Oran, Algeria, 18
    Gulf of Valencia-Catalan coast, Spain, 17
    West Nile Delta, Egypt, 19
  communities
    economic impacts, 129
    energy, 130
    human health and well being, 129–130
    tourism, 130–131
  objectives, 7–8
  rural region
    Apulia, Italy, 15
    Judean Foothills, Israel, 16
    Tel Hadya, Syria, 15–16
    Tuscany, Italy, 14–15
  urban region
    Alexandria (Egypt), 13
    Athens (Greece), 12
    Beirut (Lebanon), 12–13
  water resources, 127

P

Physical indicators. *See* Indicators

R

Regional Assessment of Climate Change in the Mediterranean (RACCM), 5–6
Rural zone, climate impact
  Apulia
    location, 15
    wine production, 77
  biogeophysical and social vulnerabilities
    CIRCE rural case study locations, 82
    deforestation, 85
    grain production, 81, 83
    tree stem volume, 84–85
  case study summary, 206
  climate hazards
    CIRCE rural case study areas, 79
    precipitation and evaporation, 78–79
    rainfall, 80
    temperature, 79–81
  climate related impacts
    agriculture, 75, 77
    carbon sequestration, Judean foothills, 76
    ecosystem, 75
    forestry, 75
    freshwater resources, 75
    natural vegetation types, 77
    seed quality, 77–78
    soil, 75
    tourism, 75, 78
    water quality, 77
  economic implications, 73
  hazards and vulnerabilities, 74
  Judean foothills
    carbon sequestration, 76
    location, 16
  rainfall, 73
  Tel Hadya
    groundwater levels, 83, 84
    location, 15–16
  Tuscany
    biogeophysical and social vulnerabilities, 82
    climate, biogeophysical and social vulnerability indicators, 87–88
    climate records, 83
    economy, 86
    location, 14–15
    water availability, 86

S

Social indicators. *See also* Indicators
  challenges
    climate variability, 57
    complex type, 57
    data, 57
    regional/local level, 58–59
  themes
    coastal case-study, 56–57
    rural case-study, 56
    urban case-study, 56
Social vulnerabilities, climate impact. *See also* Vulnerability assessment
  coastal zones, 97–101
  rural region, 80–88
  urban region, 70–73
Socioeconomic projections. *See* Biogeophysical impact
Stakeholder involvement
  barriers, 34–35
  checklist, 35–36
  contribution of, case study
    adaptation strategies, 33–34
    conceptual framework and indicators, 31
    data and knowledge, 31–32
    policy recommendations, 34
    risk assessment and management, 33
    thresholds/limits identification, 32
  level
    CIRCE case study, 27
    dissemination strategies, 28
    ladder, 26, 27
  objectives
    bottom-up approach, 28
    tools and techniques, 29
    top-down approach, 28
    vulnerability, 29
    world café, 30

T

Tel Hadya
  groundwater levels, 83, 84
  location, 15–16
Temperature level
  in Athens, 67
  climate projections, 109–113
  coastal zones, sea surface, 92
  rural zone, 79–81
  sea surface temperature (SST), 120–121
  urban region, climate impact assessments, 71, 73

Index   225

Tourism. *See* Adaptation; Climate impact
    assessments
Tuscany
    biogeophysical and social vulnerabilities,
        82, 87–88
    climate records, 83
    economy, 86
    evapotranspiration, 49
    fires/area burnt, 48, 49
    location, 14–15
    water availability, 86

U
Urban Heat Island (UHI), 63–65
Urban region, climate impact assessments
    Alexandria (Egypt), 13
    Athens
        biogeophysical and social vulnerability
            indicators, 68
        daily energy consumption and daily
            mean air temperature, 67
        location, 12
    Beirut
        heat stress and water pollution, 69
        location, 12–13
    biogeophysical and social vulnerabilities
        air pollution, 71
        annual rainfall, 71
        CIRCE urban case study locations, 72
        levels, 70
        temperature level, 71, 73
    case studies, 63
    climate hazards, 69–70
    climate related impacts
        biogeophysical and social vulnerability
            indicators, 68
        hazards and vulnerabilities, 65–66
        heat stress and water pollution,
            Beirut, 69
        human health, 68
    urban heat island (UHI), 63
    urbanization, 63

V
Vulnerability assessment
    case-study work, 206
    definition, 10
    integrated
        interactive stakeholder
            approach, 49
        linkages diagram, 49–50
        potential indicators, 48
    social
        coastal zones, 97–101
        rural region, 80–88
        urban region, 70–73
    stakeholder involvement, 29

W
Water pollution, 69
West Nile Delta, Egypt, 19

Printed by Publishers' Graphics LLC
DBT130910.15.14.99